IMISCOE Research Series

This series is the official book series of IMISCOE, the largest network of excellence on migration and diversity in the world. It comprises publications which present empirical and theoretical research on different aspects of international migration. The authors are all specialists, and the publications a rich source of information for researchers and others involved in international migration studies. The series is published under the editorial supervision of the IMISCOE Editorial Committee which includes leading scholars from all over Europe. The series, which contains more than eighty titles already, is internationally peer reviewed which ensures that the book published in this series continue to present excellent academic standards and scholarly quality. Most of the books are available open access.

Nick Dines

Migration and European Cities

A Multidimensional Perspective

Springer

Nick Dines
Department of Sociology and Social Research
University of Milano-Bicocca
Milan, Italy

ISSN 2364-4087 ISSN 2364-4095 (electronic)
IMISCOE Research Series
ISBN 978-3-032-15378-4 ISBN 978-3-032-15379-1 (eBook)
https://doi.org/10.1007/978-3-032-15379-1

This Springer imprint is published by the registered company Springer Nature Switzerland AG
The registered company address is: Gewerbestrasse 11, 6330 Cham, Switzerland

Contents

1 Introduction ... 1
1.1 Where to Begin? A 'Conventional' Snapshot ... 1
1.2 Working Definitions ... 4
1.3 The Migration-City Nexus in a European Perspective ... 6
1.3.1 Migration *to* Cities, Migration *in* Cities, Migration *and* Cities ... 6
1.3.2 Towards a Multidimensional and Context-Sensitive Approach ... 8
1.4 The Ebb and Flow of Research on Migration and Cities ... 10
1.4.1 A Historical Overview ... 10
1.4.2 The Return of the City? ... 13
1.5 Organization of the Book ... 15
References ... 16

2 Historical Perspectives of Migration and Urbanization in Europe ... 21
2.1 Introduction ... 21
2.2 Migration and Urban Centres in Early Modern Europe (1500–1800) ... 24
2.3 European Cities and the Trade of Enslaved People ... 28
2.4 Migration and Cities in the 'Industrial Age' (1800–1939) ... 29
2.5 Colonial Conquest and Mass Emigration: Europeans in Non-European Cities ... 34
2.6 International and Internal Migrations and Urban Change After 1945 ... 37
2.7 Conclusion ... 42
References ... 43

3 Migration and the Transformation of Urban Economies and Labour Markets ... 49
3.1 Introduction ... 49
3.2 Migrant Labour Markets in European Global Cities ... 52

3.3 Below the Top-Tier: Migrant Labour in Less Powerful Cities 55
3.4 Gendered Migration and Reproductive Labour 58
3.5 Non-standard Employment in Informal and Digital Gig Economies 60
3.6 Migrant Entrepreneurship and Urban Development 63
3.7 Conclusion 67
References 68

4 Arrival, Settlement and the Remaking of European Urban Life 75
4.1 Introduction 75
4.2 Arrival: Housing Opportunities and Barriers 78
4.3 Migrant Urban Settlement Processes 81
4.3.1 Some Preliminary Remarks on Migrant Settlement Literature 82
4.3.2 Settlement Pathways in Two Cities: Bangladeshis in London and Rome 84
4.3.3 Residential Segregation 86
4.3.4 The Politicization of Segregation: No-Go Areas, Ghettos and Territorial Stigma 89
4.3.5 Contested Neighbourhood Identities, Residential Mobility and Transnationalism 91
4.3.6 The Intimate Sphere of Settlement: Migrant Homemaking in the City 93
4.4 Migrant Urbanisms: Using and Remaking the City 95
4.4.1 Open Public Space 96
4.4.2 Social Infrastructures and Religious Spaces 98
4.5 Living with Difference: Conviviality, Racism and Conflict 101
4.6 Conclusion 105
References 106

5 Migration and Urban Cultures: Representations, Practices and Consumption 117
5.1 Introduction 117
5.2 Migrant Literatures and the City 121
5.3 Migration-Related Carnivals and Festivals 125
5.3.1 Notting Hill Carnival 126
5.3.2 Cologne Carnival 127
5.3.3 Vietnamese Tết Festivals in Warsaw 128
5.3.4 Carnivals and the Migration-City Nexus 129
5.4 Migration, Cities and Cultural Heritage 130
5.4.1 Migration in City Museums 131
5.4.2 Everyday Traces of Migration Heritage and Urban Tourism 133
5.5 Conclusion 136
References 137

6 The Urban Dimension of Migration Policies and Politics 141
6.1 Introduction 141
6.2 Theoretical Debates on Urban Citizenship 144
6.3 Migration Policymaking and Cities 146
6.3.1 What's New? Historicizing Local Migration Policies 147
6.3.2 Local Migration Policies Between 'Integration' and 'Diversity' 149
6.3.3 Recent Research on Local Migration Governance 151
6.4 Cities of Refuge, Sanctuary and Solidarity: Local Policy Actors Take a Stand 153
6.5 Migrant Politics in Cities 155
6.6 The Urban Scale of Anti-migration Politics 157
6.7 Conclusion 160
References 160

7 Conclusion 167
7.1 Critical European Perspectives on the Migration-City Nexus 167
7.2 Political Ruptures, Global Pandemic and War 168
7.3 Some Pointers for Future Research 171
References 173

Chapter 1
Introduction

1.1 Where to Begin? A 'Conventional' Snapshot

This book explores the multifaceted relationship between migration and European cities. Its two main objectives are to develop a critical understanding of the different interconnections between migration and cities, and to provide readers with a guide to navigating the extensive literature that has examined related aspects. The relationship between migration and cities is deeply implicated in the transformation of contemporary societies, but despite coming under increased scrutiny in recent years, it has not always been the focus of sustained reflection. More often than not, it is evoked indirectly through common motifs such as the globally-networked metropolis and the 'multi-ethnic' neighbourhood, the hyper-mobile professional and the low-paid service worker, 'cultural diversity' and residential segregation. The ways in which migration and cities are linked are, however, potentially limitless, while the meaning of these same motifs can vary significantly from place to place and according to who uses them. Moreover, migration and cities are each the focus of well-established fields of research—'urban studies' and 'migration studies'—that cut across the social sciences and humanities and encompass an overwhelming array of theoretical, empirical and methodological perspectives. A book on such a far-reaching topic therefore faces the daunting task of deciding where to begin.

A standard starting point, especially in the global reports that have tackled the issue over the last decade (see, e.g., IOM, 2015; WEF, 2017), is to recite the demographic truism that we live in an increasingly urban world (IOM, 2015: 1). Observers often proceed to make the equally self-evident observation that 'migrants tend to concentrate in cities' (ibid.: 2). Consequently, migrants are 'driving much of the increase in urbanization, making cities much more diverse places in which to live' (ibid.: 1). Major cities around the globe soon get mentioned—Sydney, London, New York, Dubai—where foreign-born residents represent at least a third of the population (IOM, 2015: 1; WEF, 2017: 14).

N. Dines, *Migration and European Cities*, IMISCOE Research Series,
https://doi.org/10.1007/978-3-032-15379-1_1

Let us now adopt the same routine approach to introduce the European context and see where it leads us. Europe has continued to urbanize, albeit at a slower rate than the rest of the world, with currently over 70% of the continent's population residing in an urban area (Our World in Data, 2024). As in other parts of the world, many of Europe's principal cities and their surrounding metropolitan regions are home to high numbers of international migrants. According to data published by Eurostat in 2016,[1] London had by far the largest foreign-born population of any European city, with almost 3 million people, more than half of whom arrived after 1991. Greater Paris came second with 1.7 million, while Berlin, Madrid, Barcelona, Vienna and Brussels all had more than 500,000 foreign-born residents. Indeed, many of Europe's major cities were growing again after decades of population decline, thanks largely to international migration. On the basis of these figures alone, it is not surprising that thinking and writing about international migration is often associated with cities, especially large and economically powerful ones.

However, it is worth noting that during the same period, 57.5 per cent of Europe's urban population was concentrated in towns and cities with fewer than 300,000 inhabitants (Our World in Data, 2018). In fact, the overall picture of where international migrants live becomes more complicated when urban areas are ranked according to their *share* of foreign-born residents (Fig. 1.1). For example, the three Spanish cities with the highest proportions of foreign-born residents—Torrevieja, Fuengirola and Benidorm—are all medium-sized urban centres on the Mediterranean coast that have seen the arrival of significant numbers of northern European lifestyle migrants in recent decades. In Estonia, the small city of Narva in the north-eastern corner of the country tops the list, with 43.8 per cent of its population of foreign origin, reflecting the repopulation of the city by Russian workers following the Second World War and the redrawing of international borders following Estonia's independence in 1991. The top-ranking city in Italy—not included in the Eurostat publication (see footnote 1)—is Prato, where in 2024 a quarter of its 200,000 inhabitants were foreign nationals (https://demo.istat.it/), roughly two-thirds of whom were Chinese, a group closely linked to the restructuring of the local textile industry.

Such statistics appear to confirm the presence of significant migrant populations in a wide range of cities. They tell us little else. Any discussion of the connections between migration and cities based on general population data and fixed areal units is inevitably going to be very limited. Urbanization extends far beyond administrative boundaries, while the 'cities' listed are themselves highly differentiated places. For example, the districts of South Kensington, Notting Hill and North Kensington in west London illustrate the stark differences that exist within a few hundred metres: the first is home to the globetrotting super-rich; the second, a former hub of post-war

[1] The Eurostat data presented here are for illustrative purposes only and are not intended to provide an up-to-date statistical overview. At the time of writing, the 2016 publication was the most recent go-to source for detailed comparative data on foreign-born populations in European cities. As with all urban and migration-related statistics, it is important to be aware of its limitations: it does not include major European cities that are not part of the European Union (EU) or the European Free Trade Association (EFTA), such as Moscow or Istanbul, nor does it include EU Member States such as Italy that do not collect data on foreign-born residents.

Country	City with highest % of foreign-born residents		City with second highest % of foreign-born residents		City with third highest % of foreign-born residents	
Austria	Vienna	31.4	Salzburg	26.8	Innsbruck	23.9
Belgium	Brussels	44.4	Antwerp	28.2	Liège	25.0
Bulgaria	Blagoevgrad	2.8	Sofia	2.5	Varna	2.4
Czech Republic	Prague	15.1	Karlovy Vary	14.3	Karvina	10.4
Croatia	Slavonski Brod	28.9	Zagreb	17.7	Rijeka	15.9
Estonia	Narva	43.8	Tallinn	20.3	Tartu	8.7
Finland	Helsinki	11.9	Vantaa	11.9	Espoo	11.6
France	CA Val de France	38.0	CA des Lacs de l'Essonne	29.6	Argenteuil-Bezons	29.1
Germany	Offenbach am Main	40.3	Frankfurt am Main	34.6	Villingen-Schwenningen	32.0
Greece	Athens	14.7	Thessaloniki	14.3	Ioannina and Kavala	10.7
Hungary	Budapest	7.3	Szeged	7.3	Debrecen	4.9
Ireland	Galway	26.1	Dublin	21.1	Waterford	17.4
Latvia	Daugavpils	22.8	Riga	20.9	Liepaja	20.5
Lithuania	Klaipeda	12.9	Vilnius	10.7	Siauliai	4.6
Norway	Oslo	24.7	Stavanger	18.0	Kristiansand	13.9
Portugal	Amadora	17.9	Sintra	17.2	Odivelas	15.6
Slovenia	Ljubljana	19.0	Maribor	14.2	-----	
Spain	Torrevieja	48.7	Fuengirola	38.5	Benidorm	32.8
Sweden	Malmö	30.5	Gothenburg	22.8	Stockholm	22.6
Switzerland	Geneva	51.3	Lausanne	46.8	Lugano	46.2
United Kingdom	Slough	39.0	London	36.7	Leicester	33.6

Fig. 1.1 Urban settlements of more than 50,000 inhabitants in EU/EFTA member states with the highest share of foreign-born residents in total population, 2011–2014. *Source* EUROSTAT (2016)

labour migration, is now largely gentrified and counts many migrant professionals among its residents, while the third was the site of the 2017 Grenfell Tower fire, in which undocumented migrants and asylum seekers were among the many who perished. It is also important to remember that migration to urban areas does not necessarily involve crossing international borders. Throughout history, it has also been closely associated with internal movements from rural and poorer regions. Moreover, territorial changes can turn local inhabitants into 'foreign-born residents', as in the case of Narva, while city dwellers who have never been abroad can still be defined as 'immigrants' or 'foreigners' on the basis of their ancestral origins or lack of formal citizenship, as in Prato and other Italian cities. In other words, commencing a book on migration and cities by listing Europe's top 'migrant cities' might offer an intriguing preliminary overview that perhaps reflects the contemporary penchant for urban benchmarking (Acuto et al., 2021). Ultimately, however, this snapshot is most effective in raising questions about what gets left out of the picture.

The following chapters explore how and why migration and cities are intertwined. While numerous studies have examined migration in urban contexts, relatively few have probed the broader links between migration and cities, and most of these have tended to focus on specific dimensions such as migrant entrepreneurship or local governance. Migration does not just transform the demographic composition of cities: it shapes and alters, among other things, their economic development, social relations, cultural practices, political life, policy agendas, urban morphology and sensory landscapes. Conversely, the distinctive features, histories and networks of cities can influence the types of migration and the conditions under which newcomers settle in a place and interact with local society. Viewing migration through the lens of the city and, vice versa, the city through the lens of migration invites us to consider urban and migratory phenomena in new and productive ways. Indeed, the city is increasingly used by scholars as an alternative scale to the nation-state for addressing the socio-economic implications of migration (Glick Schiller & Çağlar, 2009). From the opposite perspective, migration has been used to challenge territorially bounded ideas about the politics, economies and identities of cities (Nicholls & Uitermark, 2017; Whitehead et al., 2017).

The remainder of the introduction is structured as follows. First, I provide working definitions of migration/migrant and cities/urban in order to establish a general interpretative framework. I then outline the approach and aims of the book, explaining the relevance of nexus thinking for addressing the multiple linkages between migration and cities and the reasons for the comparative focus on European urban contexts. I continue with a brief review of over a century of research on migration and cities in order to situate the recent surge of interest, before finally outlining the content of the book's five thematic chapters.

1.2 Working Definitions

So far, I have used the terms 'migration' and 'city' as if they alluded to coherent, self-evident phenomena. This is, of course, far from the case. Both are complex terms, referring as much to processes, scales and relations as to particular types of people or places. Their connotative baggage and associated vocabularies shift over time and across different contexts. Given the limited space here, and the book's ongoing reflection on how their meanings change in relation to each other, I will limit myself to a few key points to orient the reader.

The word 'migration' is used in this book to refer to the various types of human movement—international, internal, labour, professional, forced, lifestyle, undocumented, etc.—that result in the temporary or permanent settlement of people in cities. Similarly, I choose to use 'migrant' in a broad sense to denote the very different kinds of people who come to cities from elsewhere. For stylistic reasons, unless otherwise stated, 'migration' and 'migrant' are sometimes used interchangeably with 'movement' and 'human mobility' on the one hand, and 'newcomers' and 'recent arrivals', on the other. The term 'migration background' is instead used (sparingly) to refer to people who have a recent history of migration in their immediate family. Words such as 'non-migrant' and 'native' are employed to indicate existing inhabitants of a city, albeit on the premise that their own ancestral histories are likely to be characterized by migration and that this population is not coterminous with local social and economic systems.

None of the above terms are satisfactory, and indeed all will be qualified and problematized in the course of the book. Migration and migrant are not neutral descriptors of human mobility. There are political, legal and social mechanisms that govern how some people move more easily than others and why some people are more likely to be categorized as 'migrants.' British residents in Torrevieja or Benidorm are typically called 'ex-pats' rather than 'lifestyle migrants', while there is an unspoken assumption in much public discourse that 'migrant' indicates a foreigner from a low-income country who is often, if not always, a person of colour. A broad starting definition of migration thus serves to bring into view different forms of movement that overlap in urban contexts, but also to illuminate the conditions under which this mobility takes place and the differentiated processes that unfold after arrival and, crucially, to reflect on the ways in which knowledge about migration and cities is produced.

To do all this requires grappling with two interrelated tensions. The first comes in the form of a question: when do people in the city stop being 'migrants'? At what point do they become someone else—citizens, residents, members of 'ethnic minorities', etc.—and under what circumstances might their migrancy be reaffirmed? Thinking along these lines keeps questions such as racism and socio-economic inequality in the foreground, but also compels us to reflect carefully on what constitutes research on migration and cities and what happens when proxy terms such as 'diversity' and 'minorities' enter the discussion. The second tension concerns the distinction that is routinely made between 'migrants' and 'non-migrants', the latter often clumsily subsumed within notions of the 'local' and, in Europe at least, entangled with ideas about whiteness. In recent years, a number of scholars have confronted the politically constructed nature of this distinction by calling for the 'de-migranticization' of migration research as a discrete object of study, insisting that human mobility should instead be conceived as an integral force in the restructuring of contemporary capitalist society (Anderson, 2019; Dahinden, 2016). In urban contexts, this means understanding not only how different migrants face particular conditions—segmented labour markets, discrimination, etc.—but also how they are implicated in the same urban dynamics along with the wider population (Çağlar & Glick Schiller, 2018).

The word 'city' is used alongside related adjectives and nouns such as 'urban' and 'urbanization'. The simple question 'what is a city?' is notoriously difficult to answer. Any attempt is invariably torn between taken-for-granted assumptions, subjective experience, empirical variation and critical theory, rendering a singular definition futile. A key distinction lies between seeing the city as a particular place characterized by size, density and heterogeneity that affords opportunities for employment, consumption and socialization, and viewing the city as the outcome of relations, functions and flows that facilitate the production and reproduction of capital and people. The first perspective reflects conventional sociological ideas about cityness and urban life (Wirth, 1938). The second perspective resonates with the concepts of 'urbanization', understood not simply as the physical and demographic growth of cities but as the social and economic reorganization of urban space to enable capital accumulation (Harvey, 1985), and 'urban scale', which refers not to a fixed position in a hierarchy of nested scales but how the urban is differently positioned in broader, evolving spatial relations of power (Brenner, 2019).

We can think through this distinction in relation to migration. The first perspective captures migrants' experiences of the city as a distinct place and desired destination, but it also risks reifying the city as a pre-given, self-enclosed entity. The second perspective potentially allows for a more dynamic understanding of the urban world—one that is interconnected with other places and scales and can acquire multiple and unexpected forms—but it likewise runs the risk of reducing the city/urban, and migration with it, to a set of abstract processes. Some of the most penetrating analysis of migration and cities engages with these underlying tensions. A powerful example of synthesis between 'city-as-place' and 'city-as-process' is *Migrant City*, a book co-authored by two sociologists Les Back and Shamser Sinha and their migrant research participants, which explores the stimuli and struggles of living in the hyperconnected and convivial but also hostile environment of contemporary London (Back et al., 2018). The following chapters also seek to address these tensions by exploring how migration is closely bound up with processes of urbanization and how it contributes to shape and define the actual places that people continue to call 'cities.'

1.3 The Migration-City Nexus in a European Perspective

1.3.1 Migration to *Cities, Migration* in *Cities, Migration* and *Cities*

This book draws attention to the plurality of theoretical, methodological and empirical perspectives in the literature on migration and cities. In order to begin to grasp the myriad issues at stake, three basic approaches can be identified.

The first approach—'migration *to* cities'—focuses on the act of movement. This approach has long preoccupied historians and demographers, who have studied various flows to cities and their influence on urban growth (De Vries, 1984). Today, scholars rarely limit their scrutiny to one-way journeys, but are interested in exploring the influence of border regimes on people's circuitous routes, as well as the ongoing mobility *from* and *between* cities and how this contributes to transnational connections (Eade & Smith, 2008). Research has considered a range of issues, including the different migration systems that facilitate population movements to cities and the economic and cultural networks that emerge in their wake.

The second approach can be described as 'migration *in* cities.' Rather than a destination or a node in a network, here the city functions as a setting for investigating migration after arrival. This approach, which covers a wide range of issues from residential patterns to group formation, characterizes the vast bulk of research on migration and cities. It has resulted in fine-grained ethnographic studies and meticulous quantitative analyses of the ways in which migrants' lives are shaped by urban contexts. But it has also produced work in which the city is conceived as little more than a container for action or as a microcosm of the national situation. Either way, the primary concern of the 'migration *in* cities' approach tends to be the multifarious domains of migrant experience rather than a general reflection on the city itself.

The third approach—'migration *and* cities'—switches our attention to the synergies between migratory and urban processes. Urban localities are seen to play a prominent role in migration theory building, while migration is recognized as directly implicated in urban development. This approach characterizes the work of a number of historians and social scientists working at the intersection of urban and migration research (Arbaci, 2019; De Munck & Winter, 2012). In the field of migration studies, it is most closely associated with the work of anthropologists Nina Glick Schiller and Ayse Çağlar, who explore, among other things, the impact of international migration on contemporary urban restructuring processes (Çağlar & Glick Schiller, 2018; Glick Schiller & Çağlar, 2009).

While the latter approach underpins much of the recent theoretical work, any understanding of the relationship between migration *and* cities cannot be built without an analysis of flows *to* and processes *in* cities. Indeed, studies often necessarily combine aspects of all three approaches. Research on transnationalism, for example, considers the movements to and from cities, how these reconfigure settlement pathways in different localities, but also how these situate migrants 'fully within the forces that are constitutive of the *urban*' (Glick Schiller, 2012: 37). It is important therefore to be alert to the kinds of knowledge that might be produced by emphasizing one perspective over another. For example, the answer to the elementary question 'who is a migrant?' might, depending on the perspective adopted, be someone on the move, someone who has originated from elsewhere, or someone whose impact on a city should be seen through the same analytical lens as 'native' residents.

Together, these three approaches comprise what I want to call the 'migration-city nexus.' In recent years, nexus thinking has proliferated in migration studies, particularly in policy-related contexts. Although sometimes applied superficially, it remains particularly useful for identifying the mechanisms and discourses through which migration and other processes, such as development, co-create and reinforce each other (Carling, 2017). Here, I use the idea of nexus not as 'fancy wrapping' but as 'inspiration for thinking about the interconnections that [we] might not otherwise have considered' (ibid). The migration-city nexus thus refers to the multiple links and interdependencies that exist between different types of migration and cities, and serves as a reminder that this complexity cannot be neatly captured by a single overarching theory, catch-all concept or exemplary case.

1.3.2 *Towards a Multidimensional and Context-Sensitive Approach*

The book is organized around two guiding principles. First, as indicated above, the book adopts a multidimensional approach. The five main chapters focus on the historical, economic, socio-spatial, cultural and political/policy dimensions of the migration-city nexus. This means engaging with different thematic and disciplinary perspectives that are not usually brought into conversation. It also means recognizing how broader social divisions such as gender, race and class intersect and inform our understanding of the migration-city nexus. For example, what role does gender play in migrant-dominated labour markets, or how might racism shape housing outcomes? This underlines the importance of seeing migration as one of a number of key processes that shape urban society rather than as an extraneous phenomenon. For this reason, I choose to use 'incorporation' to describe post-arrival processes as a more neutral, if imperfect, term in contrast to the loaded notion of 'integration' which tends to cast host society as a homogeneous whole (Schinkel, 2018) and is better understood as a particular local policy instrument (Sect. 6.3.2). The interplay between the attraction of the city as a place of possibility on the one hand, and its reality as a place characterized by unequal distribution of power and resources on the other, lies at the heart of the relationship between migration and cities, and is illuminated in different ways in each of the chapters.

Second, the book draws attention to the significant variations between and within cities. The field of urban studies has traditionally been dominated by the major cities of northern America and western Europe, which have skewed our ability to assess the heterogeneity of urban experiences across the rest of the world (Robinson, 2006; Roy, 2009). Today, Europe's cities and urban regions have relatively small populations and low growth rates compared to the rapidly expanding metropolises of Asia and Africa. Moreover, only a small fraction of the roughly one billion international and internal migrants worldwide actually move to cities in the West (WEF, 2017: 14). In other words, Europe is neither at the centre of global urbanization nor does it possess a monopoly on insights into urban migration. Thus, at one level, a study of migration and cities in Europe must carefully frame its discussion in order to counter any suggestion that what occurs here is immediately relevant elsewhere.

At the same time, a critical focus on Europe allows us to foreground the very uneven playing field in the production of knowledge about the migration-city nexus. Without underestimating Europe's involvement in the histories of colonialism and racism that continue to structure global migration regimes (Mayblin & Turner, 2021), the postcolonial challenge of viewing all cities as 'ordinary cities' (Robinson, 2006) can also be applied to Europe and is indeed necessary if we are to develop a more rigorous interpretative framework. European cities are very different in terms of their development, political traditions, class composition, relations with other places and experiences of migration.In addition to its major centres of economic, political and cultural power such as London, Paris and Berlin, Europe has many cities that have had little bearing on international debates and research agendas in either urban or migration studies (Schmiz et al., 2022). These include cities in southern Europe that experienced late or limited industrialization, 'post-socialist' cities in Central and Eastern Europe[2] that have been subjected to the double shock of market reforms and privatization since the 1990s, but also peripheral, 'disempowered' towns and cities in the continent's north-western quadrant (Çağlar & Glick Schiller, 2018; Ferenčuhová & Gentile, 2016; Leontidou, 1996). A serious analysis of the migration-city nexus in Europe therefore needs to engage with a wide range of urban contexts, especially those places outside core regions that unsettle preconceptions about what we mean when we talk about 'European cities' (Colomb & Kazepov, 2023).

This insistence on appreciating contextual differences within Europe has arisen out of my own work on the cities of Naples and London. Research on Naples over many years has included an ethnographic study of the lives of international migrants around the central railway station against the backdrop of urban regeneration in the city's historic centre (Dines, 2002, 2012). Research in London in the 2000s explored the social and health implications of different public spaces in the east of the city, particularly for different 'ethnic minority' groups (Dines, 2009; Dines & Cattell, 2006). Naples and London are extremely different in terms of their power and wealth, their relationship to global capitalism, and with respect to ideas about urbanism and human mobility. London is a post-imperial capital at the core of the world economy, long construed as one of Europe's leading cities of migration and a template for interpreting epochal changes such as globalization. Naples, on the other hand, has traditionally been deemed by outsiders and insiders alike to be a problem-riddled metropolis on the margins of both Europe and Italy. Swollen by mass rural-to-urban immigration in the early modern period, the city was later the point of departure for millions of southern Italian emigrants in the nineteenth and twentieth centuries and it was not until the 1980s that it began to experience substantial immigration from abroad.

Researching these two cities called for a careful engagement with distinct migration and urban histories, but also with the political and intellectual environments in which they were understood. At the time of my research in London in the mid-2000s, the term '(im)migration' had only recently re-entered British public debate, largely as a result of increased flows from Central and Eastern Europe following the 2004 enlargement of the European Union. Most attention on international mobility focused on asylum seekers, while government discourses on multiculturalism were framed by calls for 'community cohesion' in the wake of the disorders between 'native' and Muslim groups in northern English cities in 2001. By contrast, the question of

[2] Capital letters are used to demarcate geopolitical regions such as post-1989 Central and Eastern Europe and post-1945 Eastern Europe. The small case is instead used for general geographical areas, i.e. northern, western, central, eastern, southern Europe.

'*immigrazione straniera*' (foreign immigration) was a constant focus of attention in Italian cities throughout the 1990s and 2000s. During this period, debates about urban life in Naples were punctuated with politicized categories such as '*extracomunitari*' (persons from outside the EU) and '*clandestini*' (irregular migrants) and were interwoven with broader issues of '*sicurezza*' (security) and '*cittadinanza*' (citizenship) that previously had held little relevance in public discourse about the city (Dines, 2012: 185–194).

Analyzing the migration-city nexus in Europe therefore requires us to 'provincialize' European experiences in relation to the rest of the world, but equally obliges us to be mindful of the continent's internal differences. This book brings into comparative relief a variety of cities, from the heartlands of migration and urban studies in north-western Europe but also from southern and eastern Europe, where international migration is often a more recent historical phenomenon and where related urban dynamics such as segregation have unfolded in particular and unpredictable ways.

1.4 The Ebb and Flow of Research on Migration and Cities

1.4.1 A Historical Overview

The last decade has seen a surge in research on the relationship between migration and cities, reflected in the plethora of monographs, edited volumes, policy reports, special issues, international projects and conferences devoted to the topic (e.g. Çağlar & Glick Schiller, 2018; De Shalit, 2018; Fong & Berry, 2017; Kasinitz & Martiniello, 2019; Pisarevskaya & Scholten, 2025; Triandafyllidou et al., 2024). While this has led some scholars to declare a 'local turn' in migration studies (Zapata-Barrero et al., 2017), it is important to acknowledge that reflections on the migration-city nexus have existed for as long as there has been an interest in cities and human movement. In the nineteenth and early twentieth centuries, the links between these two processes and their outcomes were often central to analyses of modern society. Friedrich Engels famously included a chapter on Irish migrants in Manchester in his 1845 investigation *The Condition of the Working Class in England*, which highlighted the fundamental role of this unskilled labour force or 'reserve army' in the development of industrial capitalism (Engels, 1987 [1887]). Ernst Georg Ravenstein's 'Laws of Migration' (1885), widely regarded as the first systematic study of migration, used census data to analyze mass movement towards British cities. Human mobility and the city were key themes in early sociological inquiry. Georg Simmel's reflections on early twentieth-century Berlin pointed, among other things, to the capacity of the modern metropolis and its inhabitants to accommodate the heterogeneity generated by mass immigration and to foster the conditions for a new cosmopolitan consciousness. Simmel's discussion of 'the stranger' as someone who 'comes today and stays tomorrow', was not simply embodied in an identifiable figure such as the non-native trader or the wandering Jew, but in a socio-spatial order characterized by the confluence of distance and proximity, indifference and engagement (Simmel, 1950a [1903], 1950b [1908]).

Immigration was also a foundational issue for the urban sociology that flourished at the University of Chicago during the interwar period. Attention centred on the rapid growth and transformation of Chicago, which had been driven by massive waves of migration, first of Europeans and later of African Americans from the US South (Park et al., 1967 [1925]; Thomas & Znaniecki, 1996 [1918–1920]; Wirth, 1938). In particular, the ecological perspective, which explained urban growth and residential distribution as an 'organic' process fuelled by a Darwinian competition for space and the regulatory role of land values, can be seen as an early attempt to theorize the relationship between migration and cities. The approach is famously captured in Burgess's concentric zone model, which mapped migrant neighbourhoods in Chicago and described the displacement and replacement of different groups across the city.

The Chicago School would have a lasting influence on subsequent generations of urban researchers in the West, and many of its propositions continue to resonate in concepts such as 'assimilation' and 'integration.' From the 1960s onwards its work was heavily criticized for ignoring the role of racism in urbanization, and scholars working elsewhere in the world disproved the idea that the midwestern metropolis could serve as a universal template for comprehending urban phenomena (Robinson, 2006). Nevertheless, the rich body of empirical research produced by Chicago sociologists remains one of the most sustained investigations into the migration-city nexus in the history of Western social science.

In the post-war period, a number of social scientists and public intellectuals produced seminal studies of international and internal labour migration in various European cities (Berger & Mohr, 1975; Candel, 1964; Fofi, 1964; Karpat, 1976; Simic, 1973). In different ways, these works highlighted the impact of migrant labour on urban economies and the reconfiguration of socio-spatial relations in cities. Some insights from these accounts, such as the racialization of internal migrants or the formation of political alliances between locals and newcomers, are largely overlooked in today's standard migration canon, with the exception of John Berger and Jean Mohr's extraordinary textual and visual account of Mediterranean guest workers in German and Swiss cities, which continues to be a source of inspiration for researchers (Back et al., 2018). During the same period, migrant writers, many of them women, explored the urban dimensions of arrival and settlement in their literary work, leaving us with a rich source of historical and sociological material (Sect. 5.2). In the field of history, from the 1970s onwards, scholars began to rethink the intricate web of urban migrations across Europe in the early modern and modern periods and challenged commonplaces about mobility and sedentariness in the past (De Vries, 1984; Tilly, 1978).

Despite this significant and pioneering interest in cities, the urban dimension tended to be eclipsed by the nation in the work of those scholars associated with the nascent field of migration studies. For example, Stephen Castles and Godula Kosack's important study *Immigrant Workers and Class Structure in Western Europe* (1973) considered local housing conditions and labour markets in cities such as

London, Paris and Zurich but framed its findings in terms of national rather than urban differences. As Glick Schiller and Çağlar note, researchers 'often glided seamlessly from a narrative about a particular city to generalities about migration in an entire nation-state' (2009: 182). When cities were not used to generalize about national policies and processes, they were sometimes overlooked altogether. Indeed, it is worth noting that the most influential textbook on migration in English—*The Age of Migration*, first published in 1993 and currently in its sixth edition (De Haas et al., 2019)—focuses on regional, national and global perspectives, while the urban scale has always remained marginal.

With the introduction of immigration restrictions and the closure of mass recruitment programmes in the early 1970s, general interest in labour migration also waned. The emergence of cultural studies and critical race studies, especially in the UK, shifted attention away from migration to the experiences of new minority groups and the rise of multicultural urban societies, epitomized by the work produced at the Centre for Contemporary Cultural Studies at the University of Birmingham in the UK (CCCS, 1982; Hall et al., 1978). Nevertheless, ground-breaking studies of racism, identity and coloniality, particularly in the work of black scholars Stuart Hall and Paul Gilroy, have inspired more recent discussions of migration-related issues such as urban conviviality (Sect. 4.5) and cultural production in cities (Sect. 5.1). Meanwhile, the radical renewal of urban theory by European Marxist philosophers, geographers and sociologists such as Henri Lefebvre, David Harvey and Manuel Castells marked a decisive shift away from the ecological approaches of the Chicago School towards an analysis of the relationship between capitalism and urban development, but also largely steered attention away from human mobility towards a focus on circuits of capital (Harvey, 1985; Lefebvre, 1970) and social movements and collective consumption (Castells, 1972). Apart from some interest in the political mobilization of immigrant workers (Castells, 1975), the general question of migration faded into the background. As Glick Schiller observes, 'migrants were pictured as part of the urban industrial workforce but not theorized as constitutive of transnational processes within which cities [we]re situated' (Glick Schiller, 2012: 41).

It was not until the late 1980s that social scientists began to theoretically revisit the urban dimension of migration. Through their analysis of new command centres in an increasingly globalized economy, researchers such as Saskia Sassen (1991) now placed transnational migration at the centre of their analysis of 'global' or 'world' cities (Sect. 3.2). This research agenda soon came to dominate thinking about the economic links between migration and cities, even if the focus was limited to a small group of major metropolises and to specific high- and low-wage sectors of the labour market and the types of migration associated with them. While Sassen's initial focus on New York, London and Tokyo was expanded to include competing cities in the West, such as Paris and Amsterdam, emerging players outside the West, such as Singapore and Hong Kong, and middle-ranking metropolitan centres that served as 'gateway cities' for new international migrants, such as Vancouver and Miami (Ley & Murphy, 2001; Portes, 2000), many lower-tier cities that were nonetheless shaped by globalization and had different experiences of migration remained off the radar (Robinson, 2006).

Alongside the rise of global cities, the rapid urbanization of many regions in Asia and Africa rekindled interest in rural–urban migration, particularly in China where unprecedented levels of internal mobility from the 1990s onwards led to the mushrooming of brand new megacities, such as Shenzhen near the border with Hong Kong, but also left hundreds of millions of urban dwellers without residency rights due to the hukou registration system that legally binds people to their place of origin (Chan & Zhang, 1999). These more recent trends indicated that the urban future of the planet would no longer be determined solely by Europe or North America, but now largely lied elsewhere.

1.4.2 The Return of the City?

As the non-exhaustive overview above suggests, there is an immensely rich body of research on the migration-city nexus spanning different literatures, not all of which is written in English, that needs to be taken into account. It is important therefore to contextualize the recent acceleration of interest in cities.

First and foremost, this interest reflects the re-emergence of the urban scale as a 'strategic lens for producing critical knowledge not only about the urban condition but also about major social, economic, and cultural refigurings in our societies' (Sassen, 2005, xvii). With the intensification of globalization and economic restructuring since the late twentieth century, cities—some more than others—have become drivers of capital accumulation, economic growth and uneven development and have acquired greater leverage within governance structures and political decision making. In the last two decades, scholars, but also policy advisors and journalists galvanized by the transformative power of cities (Barber, 2013; Saunders, 2010), have increasingly turned their attention to how migration is implicated in such processes. Some have seen this shift in focus as an opportunity for a crucial theoretical and methodological rejuvenation of the field of migration studies long characterized by ethnic groupism and methodological nationalism. This has prompted a closer dialogue between migration and urban theory, as well as a more reflexive treatment of the scales and units of analysis (Glick Schiller & Çağlar, 2009).

At the same time, much of this recent debate has drawn on a relatively narrow, almost exclusively Anglophone body of research. It has rarely engaged with scholars who are not normally associated with migration studies, but who nevertheless work on migration and cities, or with literatures in other languages linked to different migration, urban and intellectual histories, such as in southern Europe where post-war internal migration and urbanization were closely intertwined. Moreover, the debate has sometimes been accompanied by a rather cavalier dismissal of earlier studies as unsophisticated and therefore no longer worthy of consideration (ibid). While all this may serve to confirm a 'local turn' in the eyes of some migration scholars, it also runs the risk of flattening our understanding of the very processes that underpin the migration-city nexus.

As noted above, nexus thinking helps to identify the discourses generated at the intersection of migration and cities. The recent growth in interest has in fact been accompanied by a vocabulary of hotly contested terms: global city, segregation, incorporation, integration, and so on. The chapters in this book consider the contexts in which such terms have been used and disputed. To illustrate the conceptual baggage that has surrounded discussions, I want to briefly consider the word 'diversity' and its related neologism 'superdiversity', both of which have been frequently deployed over the last two decades to articulate the links between migration and cities.

Diversity – literally the co-presence of different types of people—often appears in the academic literature as a shorthand for the various changes that international migration brings to the composition of cities (e.g., Pisarevskaya & Scholten, 2025). It also refers to the ways in which this difference is understood, and is therefore very much a 'chaotic concept, prone to multiple interpretations and deployed to meet a range of diverse and sometimes contradictory agendas' (Raco, 2018: 11). It can be at once an everyday discourse tied to real-life experiences of difference and thus attentive to the vicissitudes and injustices that diversity can entail; a policy euphemism that distances itself from the politically discredited paradigm of multiculturalism by embracing a wider range of differences including physical ability, gender and sexual orientation (Matejskova & Antonsich, 2015); an increasingly brittle corporate framework for promoting creative and service economies and improving organizational performance through the recruitment of different, often previously excluded people (Raco, 2018); or the lodestar of liberal antiracists seeking to mitigate individual racial prejudice (Kundnani, 2023). In other words, 'diversity' needs to be qualified to have any critical purchase, otherwise it risks becoming a floating signifier with little analytical value.

Superdiversity, on the other hand, has more conceptual clarity and can be considered a direct corollary of the renewed interest in the migration-city nexus. Indeed, it was first coined by anthropologist Steve Vertovec in a 2007 article to describe the diversification of migration in a particular city, London. Superdiversity moves beyond the traditional emphasis on ethnic or national groups and their distinct migration patterns to highlight the multiple variables that shape migration at the urban scale in an era of intensified globalization, which has resulted in 'a condition of more mixed origins, ethnicities, languages, religions, work and living conditions, legal statuses, periods of stay and transnational connections that many cities have ever faced' (Wessendorf, 2018: 270). As will be shown, 'superdiversity' has been used in a variety of contexts. At the same time, it has not been without its critics, particularly in relation to its underestimation of the role of racism, power dynamics and social inequalities in shaping the meaning and substance of migration in cities (Back et al., 2018).

Rather than concepts, 'diversity' and 'superdiversity' may be better understood as 'keywords' in the sense suggested by Raymond Williams; that is, as historically contingent 'deeply familiar words [whose] meanings [are] inextricably bound up with the problems [they are] used to discuss' (Williams, 1983: 14–15). It is important to acknowledge that many scholars working on migration and cities avoid or reject 'diversity' and 'superdiversity' as organizing ideas, including those who identify with critical migration/border studies, but also those from disciplines such as geography where different terminologies are used. The lack of a shared vocabulary makes a scrupulous analysis of the migration-city nexus all the more important.

1.5 Organization of the Book

The book is divided into five thematic chapters that bring different disciplinary perspectives, methodological approaches and urban contexts into discussion. Each chapter highlights key research insights and debates, while also drawing attention to limitations and recurring dilemmas in the literature. The chapters can be read separately, but it is only by taking a multidimensional view of the migration-city nexus that one can fully appreciate the conceptual and analytical challenges that lurk beneath the surface of this relationship and overcome some of the perfunctory ways in which it is iterated in work. Each chapter begins with brief, deliberately thought-provoking sketches from one or two cities as an entry point into each theme.

Chapter 2 examines the migration-city nexus from a historical perspective. It sets the scene for the rest of the book by charting how migration has made and transformed European cities over the centuries. It also builds on recent calls for a more reflexive use of the past in migration studies (Reinecke & Löhr, 2024) to highlight the important role of historical research in achieving a deeper understanding of the migration-city nexus. Through a discussion of different types of city-bound migration since 1500, the chapter addresses breaks and continuities over time, divergences and convergences between different cities, and common misconceptions about the history of urban migration. In addition to focusing on migration within Europe, it takes into account the global reach of the European migration-city nexus by considering the key role played by certain cities in the forced movement of enslaved people to other continents, as well as the impact of European labour and colonial migration on cities in other parts of the world. The final section outlines the links between labour migration and urban development after the Second World War and introduces issues that will be considered in greater detail in subsequent chapters.

Chapter 3 analyzes the interconnections between international migration and urban economies and labour markets. The first part examines the crucial role played by different migrant workforces in Europe's leading global cities and in smaller cities further down the urban hierarchy. The chapter then discusses the gendered dimension of work and social reproduction and shows how such a focus reframes our understanding of the economic relationship between migration and cities. The following section turns its attention to the links between migrant labour and the development of atypical forms of employment, particularly in the informal and, more recently, digital gig economies. Finally, the chapter draws on international literature to look at the growth of migrant entrepreneurship in European cities over the last four decades and its contribution to urban development.

Chapter 4 focuses on the multiple social and spatial processes that unfold following the arrival of migrants in cities. In addition to research by migration scholars, it draws on the work of geographers, sociologists, planners and housing experts who have analyzed different situations across Europe. The chapter begins by outlining the shifting housing options for different migrant groups over recent decades. It proceeds to examine key settlement processes that have been at the centre of academic inquiry and sometimes heated political debate, including residential segregation, territorial stigmatization, residential mobility, transnationalism and migrant homemaking. The chapter then switches its focus to a discussion of the use and creation of public spaces and social infrastructures that sustain urban livelihoods. It ends with a consideration of everyday social interactions at the urban scale in relation to overlapping discussions about conviviality and racism.

Chapter 5 examines urban forms of cultural representation, practice and consumption and considers how this focus sheds new light on the migration-city nexus. While research on migration and cities is replete with references to 'multiculturalism' and 'cultural diversity', until recently there has been little interest in addressing cultural processes themselves. The chapter first looks at one particular form of individual expression—creative writing—to reflect on the ways in which migrant writers have represented cities since the Second World War. It then draws on scholarship on carnivals and festivals to examine the ways in which migration-themed cultural events are entangled with wider issues of urban change. Finally, it develops the idea of migration heritage by analysing the recent trend towards the integration of migration into museum displays and urban tourism.

Chapter 6 examines the migration-city nexus through the twin lenses of policy and politics. It begins by outlining recent theoretical debates on urban citizenship, showing how this broad concern intersects with both migration policy and migrant politics. Turning to the field of local policymaking, the chapter considers historical shifts in approaches to urban migration governance, discusses the different ways in which 'integration' and 'diversity' are managed at the urban scale, reviews recent lines of research, and addresses provisions for undocumented migrants such as sanctuary cities. The chapter proceeds with a discussion of migrants' political mobilizations in urban contexts and concludes by highlighting the overlooked role of cities in anti-migration politics.

In the Conclusion (Chap. 7), I draw together key threads from the chapters to reaffirm the importance of nexus thinking and contextual diversity, reflect on the impact of recent ruptures on the connections between migration and urban processes, and identify some promising areas for future research.

References

Acuto, M., Pejic, D., & Briggs, J. (2021). Taking city rankings seriously: Engaging with benchmarking practices in global urbanism. *International Journal of Urban and Regional Research, 45*(2), 363–376.

Anderson, B. (2019). New directions in migration studies: Towards methodological de-nationalism. *Comparative Migration Studies, 7*(36). https://doi.org/10.1186/s40878-019-0140-8
Arbaci, S. (2019). *Paradoxes of segregation: Housing systems, welfare regimes and ethnic residential change in southern European cities.* Wiley.
Back, L., Sinha, S. with Bryan, C., Baraku, V., & Yembi, M. (2018). *Migrant City.* Routledge.
Barber, B. (2013). *If mayors ruled the world: Dysfunctional nations, rising cities.* Yale University Press.
Berger, J., & Mohr, J. (1975). *A seventh man. A book of images and words about the experience of migrant workers in Europe.* Penguin.
Brenner, N. (2019). *New urban spaces: Urban theory and the scale question.* Oxford University Press.
Çağlar, A., & Glick Schiller, N. (2018). *Migrants and city-making: Dispossession, displacement and urban regeneration.* Duke University Press.
Candel, F. (1964). *Els Altres Catalans* (The other Catalans). Edicions 62.
Carling, J. (2017). Thirty-six migration nexuses, and counting. Blogspot at: https://jorgencarling.org/2017/07/31/thirty-six-migration-nexuses-and-counting/
Castells, M. (1972). *La question urbaine.* Maspero.
Castells, M. (1975). Immigrant workers and class struggles in advanced capitalism: The western European experience. *Politics & Society, 5*(1), 33–66.
Chan, K. W., & Zhang, L. (1999). The hukou system and rural-urban migration in China: Processes and changes. *The China Quarterly, 160*, 818–855.
Colomb, C., & Kazepov, Y. (2023). Comparative urban studies in Europe. In P. Le Galès & J. Robinson (Eds.), *The Routledge handbook of comparative global urban studies* (pp. 87–103). Routledge.
CCCS (Centre for Contemporary Cultural Studies). (1982). *The empire strikes back. Race and racism in 70s Britain.* Routledge
Dahinden, J. (2016). A plea for the "de-migranticization" of research on migration and integration. *Ethnic and Racial Studies, 39*(13), 2207–2225.
De Haas, H., Castles, S., & Miller, M. (2019). *The age of migration: International population movements in the modern world* (6th ed.). Red Globe Press.
De Munck, B., & Winter, A. (Eds.). (2012). *Gated communities? Regulating migration in early modern cities.* Ashgate.
De Shalit, D. (2018). *Cities and immigration: Political and moral dilemmas in the new era of migration.* Oxford University Press.
De Vries, J. (1984). *European urbanization 1500–1800.* Harvard University Press.
Dines, N. (2002). Urban renewal, immigration and contested claims to public space: The case of Piazza Garibaldi in Naples. *GeoJournal, 58*(2–3), 177–188.
Dines, N., & Cattell, V. (2006). *Public space, social relations and well-being in East London.* Policy Press.
Dines, N. (2009). The disputed place of ethnic diversity: An ethnography of the redevelopment of a street market in East London. In R. Imrie, L. Lees & M. Raco (Eds.), *Regenerating London. Governance, sustainability and community in a global city* (pp. 254–272). Routledge.
Dines, N. (2012). *Tuff city: Urban change and contested space in central Naples.* Berghahn.
Eade, J., & Smith, M. P. (2008). *Transnational ties: Cities, migrations and identities.* Transaction.
Engels, F. (1987 [1887]). *The condition of the working class in England.* Penguin.
Eurostat. (2016). Urban Europe: Statistics on cities, towns and suburbs. https://ec.europa.eu/eurostat/en/web/products-statistical-books/-/KS-01-16-691
Ferenčuhová, S., & Gentile, M. (2016). Introduction: Post-socialist cities and urban theory. *Eurasian Geography and Economics, 57*(4–5), 483–496.
Fofi, G. (1964). *L'immigrazione meridionale a Torino* (Southern immigration in Turin). Feltrinelli.
Fong, E., & Berry, B. (2017). *Immigration and the City.* Polity.

Glick Schiller, N., & Çağlar, A. (2009). Towards a comparative theory of locality in migration studies: Migrant incorporation and city scale. *Journal of Ethnic and Migration Studies, 35*(2), 177–202.

Glick Schiller, N. (2012). Transnationality, migrants and cities: A comparative approach. In A. Amelina, D. D. Nergiz, T. Faist & N. Glick Schiller (Eds.), *Beyond methodological nationalism: Research methodologies for cross-border studies* (pp. 37–55). Routledge.

Hall, S., Critcher, C., Jefferson, T., Clarke, J., & Roberts, B. (1978). *Policing the crisis: Mugging, the state, and law and order*. Macmillan.

Harvey, D. (1985). *The urbanization of capital: Studies in the history and theory of capitalist urbanization*. John Hopkins University Press.

IOM (International Organization for Migration). (2015). *World migration report 2015. Migrants and cities: New partnerships to manage mobility*. IOM.

Karpat, K. (1976). *The gecekondu: Rural migration and urbanization*. Cambridge University Press.

Kasinitz, P., & Martiniello, M. (2019). Music, migration and the city. *Ethnic and Racial Studies, 42*(6), 857–864.

Kundnani, A. (2023, June 29). There are two kinds of antiracism. Only one works, and it has nothing to do with "diversity training". *The Guardian*. https://www.theguardian.com/commentisfree/2023/jun/29/antiracism-diversity-training-liberal-antiracists-vocabulary-direct-action

Lefebvre, H. (1970). *La révolution urbaine*. Gallimard.

Leontidou, L. (1996). Alternatives to modernism in (southern) urban theory: Exploring in-between spaces. *International Journal of Urban and Regional Research, 20*(2), 178–195.

Ley, D., & Murphy, P. (2001). Immigration in gateway cities: Sydney and Vancouver in comparative perspective. *Progress in Planning, 55*(3), 119–194.

Matejskova, T., & Antonsich, M. (Eds.). (2015). *Governing through diversity: Migration societies in post-multiculturist times*. Palgrave Macmillan.

Mayblin, L., & Turner, J. (2021). *Migration studies and colonialism*. Polity Press.

Nicholls, W. J., & Uitermark, J. (2017). *Cities and social movements: Immigrant rights activism in the United States, France, and the Netherlands, 1970–2015*. Wiley.

Our World in Data. (2018). Share of urban population by agglomeration size, Europe, 1950 to 2035. https://ourworldindata.org/grapher/share-of-urban-population-by-size-of-the-agglomeration?country=~OWID_EUR

Our World in Data. (2024). Share of people living in urban areas. https://ourworldindata.org/grapher/urban-share-european-commission?focus=~OWID_EUR

Park, R. E., Burgess, E., & McKenzie, R. (1967 [1925]). *The city*. University of Chicago Press.

Pisarevskaya, A., & Scholten, P. (2025). *Cities of migration. Understanding the diversity of urban diversities in Europe*. Springer.

Portes, A. (2000). Immigration and the metropolis: Reflections on urban history. *Journal of International Migration and Integration, 1*(2), 153–175.

Raco, M. (2018). Critical urban cosmopolitanism and the governance of urban diversity in European cities. *European Urban and Regional Studies, 25*(1), 8–23.

Ravenstein, E. G. (1885). The laws of migration. *Journal of the Statistical Society of London, 48*(2), 167–227.

Reinecke, C., & Lohr, I. (2024). Beyond the present, the nation, and Europe: Three different uses of history in reflexive migration studies. *Migration Studies, 12*(3). https://doi.org/10.1093/migration/mnae023

Robinson, J. (2006). *Ordinary cities: Between modernity and development*. Routledge.

Roy, A. (2009). The 21st-century metropolis: New geographies of theory. *Regional Studies, 43*(6), 819–830.

Sassen, S. (1991). *The global city. New York, London, Tokyo*. Princeton University Press.

Sassen, S. (2005). Foreword. In Y. Kazepov (Ed.), *Cities of Europe. Changing contexts, local arrangements, and the challenge to urban cohesion* (pp. xvii–xxii). Blackwell.

Saunders, D. (2010). *Arrival city: How the largest migration in history is reshaping our world*. Heinemann.

Schinkel, W. (2018). Against "immigrant integration": For an end to neocolonial knowledge production. *Comparative Migration Studies, 6*(31). https://doi.org/10.1186/s40878-018-0095-1
Schmiz, A., Felgentreff, C., Franz, M., Paul, M., Pott, A., Räuchle, C., & Schrader, S. (2022). Cities and migration: Bibliometric evidence from a spatially biased field of knowledge production. *Geographical Review, 112*(2), 267–285.
Simic, A. (1973). *Peasant urbanites: A study of rural-urban mobility in Serbia*. Seminar Press.
Simmel, G. (1950a [1903]). The metropolis and mental life. In K. H. Wolff (Ed.), *The sociology of Georg Simmel* (pp. 409–424). Free Press.
Simmel, G. (1950b [1908]). The stranger. In K. H. Wolff (Ed.), *The sociology of Georg Simmel* (pp. 402–408). Free Press.
Thomas, W. I., & Znaniecki, F. (1996 [1918–1920]). *The polish peasant in Europe and America*. University of Illinois Press.
Tilly, C. (1978). Migration in modern European history. In W. H. McNeil & R. S. Adams (Eds.), *Human migration* (pp. 48–72). Indiana University Press.
Triandafyllidou, A., Moghadam, A., Kelly, M., & Şahin-Mencütek, Z. (Eds.). (2024). *Migration and cities: Conceptual and policy advances*. Springer.
Vertovec, S. (2007). Super-diversity and its implications. *Ethnic and Racial Studies, 30*(6), 1024–1054.
Wessendorf, S. (2018). Pathways of settlement among pioneer migrants in super-diverse London. *Journal of Ethnic and Migration Studies, 44*(2), 270–286.
Whitehead, C., Lloyd, K., Eckersley, S., & Mason, R. (Eds.). (2017). *Museums, migration and identity in Europe: Peoples, places and identities*. Routledge.
Williams, R. (1983). *Keywords: A vocabulary of culture and society*. Oxford University Press.
Wirth, L. (1938). Urbanism as a way of life. *American Journal of Sociology, 44*(1), 1–24.
WEF (World Economic Forum). (2017). *Migration and its impact on cities*. WEF.
Zapata-Barrero, R., Caponio, T., & Scholten, P. (2017). Theorizing the "local turn" in a multi-level governance framework of analysis: A case study in immigrant policies. *International Review of Administrative Sciences, 83*(2), 241–246.

Chapter 2
Historical Perspectives of Migration and Urbanization in Europe

2.1 Introduction

> In the streets one hears Russian, English, Italian, German, [Jewish], Tartar, Polish, Turkish, Greek, Armenian, Moldavian, Bulgarian, Hungarian, Dalmatian, French, Swedish, and Spanish, [and many other languages] and these are not spoken merely by passing strangers, but by the regular inhabitants. (Kohl, 1844: 420)[1]

The above passage is part of a detailed description of the Black Sea port of Odessa written by the German travel writer Johann Georg Kohl. In 1838, the year of Kohl's visit, Odessa had around 70,000 inhabitants. Street signs, Kohl tells us, were written in Russian, the national language, and Italian, the city's commercial lingua franca. French was the language of high society and of the local newspaper, the *Journal d'Odesse*, while the argot of the porters included borrowings from a variety of tongues. The major merchants were Greek, Italian and German, and the brokers, shipping agents and bankers were mostly Jewish. Meanwhile, Georgians had cornered the barber's trade, French and English tailors had set up shops along the acacia-lined streets, tobacconists and coffee houses were patronized by Turks, Greeks and Armenians, Tatar-speaking gypsy farriers and knife sharpeners hawked their services outside the bazaars, while Moldovan peasants from nearby Bessarabia supplied the fuel-starved city with wood and charcoal on a daily basis.

Today such a scenario might be described as 'cosmopolitan', 'multicultural' or even 'superdiverse'. Kohl instead chose an expression comprehensible to his contemporaries: this was 'the true Babylonish extreme' (420). The multilingual babble on the streets of Odessa, however, was not unique. 'The confusion of tongues', noted Kohl, 'prevails, more or less, all over Russia' (ibid). What made Odessa remarkable was that it had only been in existence for four decades, having been founded in 1794

[1] The English abridged version of Johann Georg Kohl's travelogue in Russia omitted the words '*Jüdisch*' and '*noch mehre andere Sprachen*', which were present in the original German text (Kohl, 1841: 56) and have been reinstated here in brackets.

N. Dines, *Migration and European Cities*, IMISCOE Research Series,
https://doi.org/10.1007/978-3-032-15379-1_2

as a new southern outpost of the Russian Empire. During the nineteenth century, merchants, peasants and dockworkers from across Europe and the Black Sea basin flocked to the port to take advantage of the relative freedoms and booming economy that had developed around the export of grain from the surrounding steppes. In the words of Kohl, the city's population was 'helped by Europe and Asia' (Kohl, 1841: 56). By 1897, Odessa had more than 400,000 inhabitants and had become Russia's third largest metropolis after St Petersburg and Moscow. What Kohl witnessed, then, was a city shaped simultaneously by multiple flows of migration and a process of rapid urbanization that was already occurring, or about to occur, in many other parts of Europe.

Since the 1970s, historians have tended to see migration as a constant feature of societies, rather than the result of exceptional events such as war or economic crisis (Moch, 2003; Tilly, 1978). In doing so, they have dismantled many of the temporal and spatial frames through which migration has often been interpreted. For example, studies of the multiple systems of movement between town and country in early modern Europe have put paid to the enduring notion that most of pre-industrial society was sedentary (Lucassen & Lucassen, 2009), while research on nineteenth-century mass migrations to Siberia, Manchuria, and across the Indian and Pacific Oceans has challenged the prominent place accorded to the transatlantic emigration of Europeans in historical accounts of global mobility (McKeown, 2004).

This chapter aims both to chart the evolution of European cities in relation to changing migratory flows, and to illustrate how a historical perspective provides important insights for addressing the relationship between migration and cities today. This introduction highlights three key ways in which historical research contributes to a more incisive understanding of the migration-city nexus.

First, historians have tended to embrace a broad definition of urban-bound mobility, encompassing temporary, seasonal, circular, short-distance and long-distance, individual and collective, internal and international movements. This approach contrasts with the strong focus on international migration that characterizes much of contemporary migration studies. In her survey of migration in Western Europe since 1650, Leslie Page Moch uses the elastic definition of a 'change in residence beyond a municipal boundary, be it a village or town [and] includes moves that last only a season as well as permanent relocations' (Moch, 2003: 6). This, she argues, allows her to better capture 'the nuances of historical change' (ibid). By adopting a dynamic understanding of migration, historians have long pointed out the limitations of rigid analytical frameworks, such as the dichotomy between 'forced' and 'economic' migrants. The Protestant merchants who moved from the Southern Netherlands to Amsterdam in the late sixteenth century were refugees insofar as they were oppressed by their Spanish Catholic rulers, but economic concerns—the collapse of Antwerp's textile industry—were just as much a factor in their decision to leave (Lucassen & Lucassen, 1997: 15). Critical reflections on how historians have conceptualized past movements and how they have been evaluated over time serve to better contextualize contemporary categorizations of migration and remind us that disputes over their meaning and scope are not unprecedented (Gabaccia, 2015; Henrich & Sampson, 2019).

Second, a historical perspective on urban mobility encourages a greater appreciation of the deep-rooted 'web of interconnections' (Hohenberg & Lees, 1995: 2) that have existed between different cities, between the city and countryside, and between cities and nation-states, thus countering any temptation to view the city as an autonomous, self-contained unit of analysis. Urbanization is underpinned by the movement of capital, goods, ideas and, above all, people: if this sounds like a truism in our current era of intensified globalization, it is important to note that this dimension has been constant over the centuries. Kohl, for example, recalls a merchant's enthusiasm for the speed of intercity correspondence from Odessa: 'A letter takes two to three days to go to Constantinople, eight to nine to St Petersburg, ten to Vienna, eleven or twelve to Hamburg, and from London an answer may be received over Warsaw or St Petersburg in from twenty-five to twenty-seven days' (Kohl, 1844: 419). Indeed, historians have pointed to Odessa's trade and migration links with places such as Istanbul, Livorno, Marseille and the Jewish-majority town of Brody as key reasons for the rapid development of the Black Sea port in the early nineteenth century (Herlihy, 1976; Zipperstein, 1985). Thus, current debates about translocal networks between different types of cities, and the empirical and conceptual transformations that they imply, take on greater substance when we are able to trace historical parallels.

Third, migration has always been, in the words of the early modern historian Jan de Vries, an 'unruly phenomenon' (De Vries, 1984: 200). In part, this refers to the fact that migration is historically difficult to pin down and measure. General population registers with details about a person's place of birth, date of arrival in a locality and previous residence, which allow researchers to trace movements, were not introduced until the mid-nineteenth century, and then only in a few countries. Migration and urban historians are necessarily reflexive about the possibilities and limitations of their sources and they have often had to devise ingenious ways of capturing migration patterns, such as using citizens registers, apprenticeship lists, marriage records or death certificates to identify the origins of individuals. At the same time, urban migration is not a clear-cut event like a birth, marriage or death. Instead, it often unfolds in multiple stages and directions, transforming social and economic relations along the way. The fundamental contribution of the historical method is precisely its ability to reveal the unpredictable effects of migration over time and to avoid retrospective accounts that effortlessly culminate in our current 'age of migration'. If migration has *made* cities by driving urbanization, it has also always *remade* them in unforeseen ways.

This chapter draws on a wide range of material, including historical surveys of European cities (Clark, 2009; Cornelissen et al., 2022; De Vries, 1984; Hohenberg & Lees, 1995; Todorov, 1983), histories of migration in different regions (Hochstadt, 1999; Lucassen, 2005; Moch, 2003), comparative studies of specific issues such as migration controls (De Munck & Winter, 2012; Greefs & Winter, 2019), as well as a plethora of city biographies that address the topic of migration in various ways. For reasons of space, the chapter limits its focus to the early modern, modern and contemporary periods. Its main aim is to identify and contextualize key types of urban migration in order to highlight breaks and continuities over time, to illustrate elements of divergence and convergence between different cities and to address some common misconceptions about the history of urban migration. Population charts for major European cities in the three periods have been included as reference points for the broader discussion about urbanization and demographic change. While the chapter focuses on urban migration within Europe, it also takes into account the global reach of the European migration-city nexus and thus considers the key role played by certain port cities in the forced movement of people to other continents, as well as the impact of European labour and colonial migration on cities in other parts of the world. Finally, the last section outlines the relationship between post-colonial, guest worker and internal migration and urban development in Europe after the Second World War, also in order to provide a general historical context for issues that will be considered in greater depth in later chapters.

2.2 Migration and Urban Centres in Early Modern Europe (1500–1800)

Until recently, the dominant image of early modern Europe was that of a largely sedentary peasant society in which population upheavals were mainly the result of religious persecution, war, famine or disease. According to available demographic information (De Vries, 1984), only a few major ports and political centres such as Paris, Naples, London, Amsterdam and Istanbul experienced significant growth (Fig. 2.1). Otherwise, the average urban settlement had around ten thousand inhabitants in 1500 and its population remained stable or grew very slowly over the next three hundred years.

The great metropolises of the time were characterized, as they are today, by their cultural and religious diversity. This was particularly the case in Amsterdam, where foreigners made up more than a third of the city's population in the early seventeenth century (van Lottum, 2007: 63). Protestant merchants from Antwerp, Sephardic Jews from Portugal and Spain and Huguenots from France—as well as thousands of labourers, domestic servants and sailors from Germany and Scandinavia—all contributed significantly to the city's commercial and financial supremacy during this period. Meanwhile, after the fall of Byzantine Constantinople in 1453, the new Ottoman rulers in Istanbul ordered the resettlement of Muslims, Christians and Jews from across the empire to repopulate the city and alleviate the severe labour shortage. In the centuries that followed, the authorities regularly moved people in and out of Istanbul to control its growth (Kasaba, 2009). Naples was different. Under Spanish rule (1503–1707) administrative elites and master artisans came to the city from all over Europe but the vast majority of newcomers came from the surrounding countryside. As a capital city, Naples enjoyed institutional connections and fiscal privileges that attracted both provincial nobles and peasants fleeing feudal oppression. The massive influx of people was accompanied by intense urbanization, transforming

	1500	1550	1600	1650	1700	1750	1800
Amsterdam	14	30	65	175	200	210	217
Antwerp	40	90	47	70	70	46	60
Barcelona	29	35	43	44	43	50	115
Berlin	12	n.d.	25	12	55	90	150
Bordeaux	20	20	40	40	50	67	88
Bristol	10	10	11	20	25	45	64
Brussels	35	40	50	69	80	60	74
Cologne	30	35	40	45	42	43	42
Copenhagen	n.d.	n.d.	n.d.	23	70	93	101
Dublin	0	5	5	17	60	90	168
Edinburgh	n.d.	n.d.	30	35	40	57	82
Gdansk	20	26	50	70	50	46	40
Genoa	60	65	71	90	80	87	91
Hamburg	14	29	40	75	70	75	100
Istanbul	200	660	700	n.d	700	625	570
Lisbon	30	98	100	130	165	148	180
Liverpool	0	0	0	0	6	22	78
London	40	80	200	400	575	675	865
Lyon	50	70	40	75	97	114	100
Madrid	0	30	49	130	110	109	167
Manchester	0	0	0	5	9	18	70
Marseille	n.d.	30	40	66	75	68	78
Milan	100	69	120	100	124	124	135
Moscow	80	75	80	n.d.	114	146	248
Naples	150	212	281	176	216	305	427
Palermo	55	70	105	129	100	118	139
Paris	100	130	220	430	510	576	581
Prague	n.d.	n.d.	n.d.	n.d.	39	59	77
Rome	55	45	105	124	138	156	163
Seville	25	65	90	60	96	66	96
Stockholm	0	0	n.d.	40	45	60	75
Turin	0	14	24	37	42	57	82
Venice	100	158	139	120	138	149	138
Vienna	20	n.d.	50	60	114	175	231
Warsaw	0	10	15	20	15	23	63

Fig. 2.1 Population (in thousands) of selected major European cities 1500–1800.
n.d.: no data, 0: less than 10,000 inhabitants.
Source De Vries (1984), Chandler (1987) for Istanbul and Moscow

Naples into one of the most populous, dense and high-rise cities in early modern Europe (Dines, 2012: 32–33).

As these three examples suggest, city-ward migration was the result of a number of overlapping factors. Some of these were general: the early modern metropolis provided a refuge for persecuted groups, but it also offered greater opportunities for professional advancement, paid work and charitable support for people arriving from nearby or further afield. Other factors were more context specific: migration was key to Ottoman plans to establish Istanbul as a new imperial capital, while it was tolerated by the Spanish viceroys in Naples as a means of maintaining control over the local nobility and easing tensions in the impoverished countryside. These flows were also socially heterogeneous. Career migration tended to be restricted to elites such as the clergy and teachers, wealthy refugees often moved in family units, whereas migrant labourers were usually of peasant stock. Moreover, in many of Europe's major cities, the majority of new arrivals were women. In seventeenth-century London, for example, it is estimated that there were 87 male migrants for every 100 women. As Leslie Page Moch writes, '[London] needed women in great numbers to produce yarn and thread, to launder clothing, to shop, to sell and prepare food, to bear and raise children, to sew clothing and serve beer—in short, to carry out particular service, productive and reproductive tasks that maintain urban life' (Moch, 2003: 56).

It would be wrong, however, to see migration as a phenomenon that only affected Europe's largest cities. On the contrary, as early modern historians are quick to remind us, it was a constant feature of all urban life (De Vries, 1984). The key point is that no urban settlement could grow naturally because of the high mortality rates of the period. Outsiders were essential to maintain the population of even the smallest towns and to replenish it in the event of war, famine or disease. Without continuous migration, most urban centres would have declined or simply ceased to exist.

Early modern Europe was therefore 'bustling with movement' (Moch, 2003: 22). Much of this movement consisted of short-distance, circular mobility between town and country. Many urban jobs in construction or textiles were seasonal or temporary, while urban dwellers were also attracted by work opportunities in rural areas, especially at harvest time. As Philip Benedict, historian of early modern France, has evocatively suggested, towns and cities were like accordions, expanding and contracting in response to the fluctuating movements of people (Benedict, 1989: 12).

At the same time, long-distance travellers were not always drawn to Europe's major cities such as Amsterdam and London. Many international migrants moved to smaller settlements that promised economic returns and employment prospects. After their expulsion from Spain in 1492, many Sephardic Jews settled in Salonica (Thessaloniki) and later in Livorno, from where they developed trading networks throughout the central and eastern Mediterranean, while Flemish weavers moved across the North Sea to the English provincial towns of Norwich and Colchester in the late sixteenth century, where they revived the local cloth trade and formed neighbourhoods with their own churches and community facilities (Goose, 1982). Even in central-eastern Europe, often assumed to be a bedrock of stability in this period due to greater feudal restrictions on mobility (Moch, 2003: 9), commercial and administrative centres such as Gdansk, Prague and Kraków were home to an array of religious, national and linguistic minorities drawn from states within the region and from further afield, including Scotland, France and the Italian peninsula (Miller, 2008).

If the pre-industrial city was a central hub for multiple types of mobility, it was also, crucially, the main place where attempts were made to regulate these movements (De Munck & Winter, 2012; Greefs & Winter, 2019). Urban authorities exercised considerable influence over who could enter and settle in cities. They had a range of control mechanisms at their disposal, from official registration at city gates and informal surveillance in guesthouses to restricting access to citizenship and issuing expulsion orders. The distinction between 'wanted' and 'unwanted' migrants fluctuated according to shifting priorities: the constant demand for outsiders to repopulate the city or boost the economy was sometimes offset by the need to reduce the burden on charitable institutions, prevent the spread of infectious diseases, and maintain social order. The granting of urban citizenship to newcomers was also a flexible process. Citizens and their families made up about half of a city's population in the Dutch Republic and German-speaking territories, and as little as 10–20 per cent in central-eastern Europe, but in most cases the vast majority of this group was born outside the city (Miller, 2008; Moch, 2003). Moreover, in order to attract and integrate skilled artisans and well-connected merchants into the social hierarchy, city authorities regularly overlooked geographical origin and linguistic and religious differences in order to offer fast-track access to guild membership and tax exemptions. In some cities, authorities attempted to govern the internal movement of minority groups as they grew in size and importance. The most notorious example is Renaissance Venice, where the world's first ghetto was established in 1516 to contain and control the well-established Jewish community. This actually followed similar measures of confinement of German and Ottoman Muslim merchants in other parts of the city, and did not prevent contemporary writers from commenting on the multicultural make-up of the city's daily life and transactions (Salzberg, 2018).

In practice, immigration controls were difficult to enforce, especially in large cities with limited policing and multiple points of entry. In addition, the divergent interests of religious, political and economic elites often meant that there was no common position on entry restrictions. Partly in response to the ineffectiveness of controls, professional police forces developed from the late seventeenth century to become the primary means of regulating and monitoring the movements of urban populations (De Munck & Winter, 2012). This period also saw the establishment of institutions such as workhouses, which were designed to capture and immobilize the growing numbers of itinerant and destitute people.

With the rise of nation-states and the intensification of centralized administrations, controls on mobility would shift from city gates and walls to external national borders. Papers that indicated local affiliation would be replaced by documents such as passports that stipulated national identity (Torpey, 2000), the distinction between insiders and outsiders would be replaced by a separation between 'citizens' and 'aliens', and the demarcation between mobility within and from outside a country would henceforth become more pronounced (Noiriel, 2019). Historians caution against reading such changes through a teleological narrative of modernity and the attendant assumption of a clear-cut rupture. Instead, they highlight how the practices of cities and nation-states were underpinned by similar concerns, such as regulating labour markets, maintaining political and social order, and managing public sentiments about immigration (De Munck & Winter, 2012; Greefs & Winter, 2019; Prak, 2018).

2.3 European Cities and the Trade of Enslaved People

A number of early modern cities became lynchpins in the management of another kind of human movement, albeit one that was entirely unfree and rarely saw the shores of Europe. The role of European cities in the history of slavery has long roots stretching back to ancient Greek and Roman times. From the Middle Ages onwards, many Mediterranean cities were involved in the two-way slave trade between Europe and North Africa and the Levant. In the fifteenth and sixteenth centuries Istanbul acted as an intermediary for up to half a million Slavic people captured by Tatars and sold to owners throughout the Ottoman Empire. Between the late sixteenth century and early nineteenth centuries, around half a million Muslim slaves lived on the Italian peninsula, mainly in coastal or southern cities (Bonazza, 2019). Around four hundred of them were used to build the gigantic Bourbon royal palace in Caserta in the late eighteenth century (ibid).

Slavery would reach its zenith with the triangular transatlantic trade of the early modern period, in which some twelve million Africans were forcibly shipped to the Americas (Blackburn, 1997: 3). In the hundred years before the formal abolition of the slave trade in Britain in 1807, three million sub-Saharan Africans alone were transported across the Atlantic on ships originating in London, Bristol and Liverpool (Morgan, 2000: 10). Other European cities involved in the global slave trade included Bordeaux and Nantes in France, Lisbon, Cadiz, Seville and Barcelona on the Iberian Peninsula, and Amsterdam. Much of this forced labour was for plantations, but it was also sometimes used to establish settlements in the colonies. A notable example is Cape Town in South Africa, where Dutch settlers imported slaves from territories bordering the Indian Ocean and used them to build the city and develop local economic activities such as the wine industry. Similarly, without the slave trade, Bristol and Liverpool would have remained small ports. Profits from the sale of slaves and slave-produced commodities such as sugar and tobacco were invested in land and lavish civic architecture, underpinning the subsequent economic development and prosperity of the two cities (Dresser, 2001; Longmore, 2007).

A small number of chattel slaves were brought back by their owners on the return voyage to British ports and either remained bound in service or were freed to add to the country's fledgling black communities. It is estimated that Britain's black population numbered 10,000 in 1770, half of whom lived in London (Killingray, 2011: 211), where figures such as Olaudah Equiano and later Robert Wedderburn would contribute to the nascent abolitionist movement and to radical working-class politics in the city (Gilroy, 1993: 12–13; Linebaugh & Rediker, 2000).

If the adage 'migration makes cities' is to have any analytical purchase across history, then the forced movement of enslaved people must be at the heart of the equation. Its impact extends beyond the slave ports of Europe. Slave-produced goods in the Americas, combined with cheap imports from the colonies, were crucial ingredients in the triumph of urban-based industrial capitalism in Europe in the nineteenth century (Blackburn, 1997; Du Bois, 1935; Lowe, 2015).

2.4 Migration and Cities in the 'Industrial Age' (1800–1939)

Nineteenth-century Europe is typically associated with the rise of urban industrial society. The image of the great, smoke-belching metropolis features prominently in accounts of the period. This, we are duly informed, was a very ambivalent city: one that promised greater freedoms, opportunities and diverse experiences, but invariably came with squalid living conditions, exploitation and violence. If in 1800 there were only 23 urban centres in Europe with more than 100,000 inhabitants, a century later there were 135 (Moch, 2003: 126) (Fig. 2.2). Cities, it seems, were now absorbing a never-ending stream of newcomers. According to a well-known hypothesis by the US geographer Wilbur Zelinsky, the nineteenth century saw a 'mobility transition' whereby 'the onset of modernization [brought] with it a great shaking loose of migrants from the countryside' (Zelinsky, 1971: 236). Such a claim echoed the views of a long line of social theorists—starting with Ferdinand Tönnies and Georg Simmel—who considered the urban concentration of the population as a hallmark of modern society and the decline of traditional ways of life.

However, most historians now reject the idea of a sudden and unprecedented exodus to cities. As we have seen, urban migration was already a structural feature of early modern Europe. Rather than the emergence of *new* forms of mobility, the nineteenth century saw a shift in the *level* of migration.[2] The surge in rural–urban migration was the combination of population growth in the countryside and a shift in the mode of production. The marked reduction in wars, plagues and famines across Europe in the eighteenth century had led to a sharp increase in a mobile, property less rural proletariat accustomed to seasonal and unstable employment between town and country. The steady shift of European capital away from agriculture and dispersed rural industries towards factory production in urban areas with a far greater demand for labour, meant that towns and cities were increasingly able to attract and retain this mobile population. In other words, people were not suddenly uprooted, but their migratory routes were directed towards more stable, if not permanent, urban jobs. Another explanation was the Europe-wide revolution in transport and communication, especially the expansion of the railways after 1850 (Lucassen & Lucassen, 2009). Travel was faster and cheaper than ever before, and across the continent previously remote towns and villages were now integrated into vast inter-city networks.

Migration was not the only driver of urbanization. Improvements in urban sanitation led to a significant decline in urban mortality rates over the course of the nineteenth century. Although there were occasional outbreaks of infectious disease, cities were no longer the death traps they had been in previous centuries. Towards the

[2] It is important to note that not all European cities experienced unprecedented levels of inbound mobility during the nineteenth century. In particular, migration rates to Amsterdam and surrounding cities were higher in the seventeenth century – when the Dutch Republic was the most economically buoyant and densely urbanized part of Europe – than they were two centuries later (Lucassen & Lucassen, 2009: 370–71).

	1800/1	1850/1	1860/1	1870/1	1880/1	1890/1	1900/1	1910/1	1920/1	1930/1
Amsterdam	217	224	244	264	326	408	511	574	642	752
Athens	12	31	41	45	63	108	111['07]	167	301	453
Barcelona	115	175	180	n.d.	346	397	533	587	710	783
Berlin	172	419	548	826	1,122	1,579	1,889	2,071	3,801['19]	4,243['33]
Birmingham	74	233	296	344	437	478	523	840	922	1,003
Bristol	64	137	154	183	207	289	339	357	377	404
Brussels[1]	66	251	281	314	421	500	599	720	685	840
Bucharest	32	120	122	142	n.d.	220	276['99]	341['12]	309['17]	631
Budapest	54	178	187['57]	202	371	506	732	880	1,185	1,006
Cologne	50	97	121	129	145	282	373	517	634['19]	757['33]
Copenhagen	101	129	155	181	235	313	401	559	561	771
Duisburg	4	9	13	31	41	59	93	229	244['19]	440['33]
Glasgow	77	357	420	522	587	658	776	1,000	1,052	1,093
Hamburg	130	132	134	240	290	324	706	931	986['19]	1,129['33]
Istanbul	570	785	n.d.	873['75]	n.d.	874	1,125	1,200	n.d.	691['27]
Kiev	23['11]	50	68['63]	71['67]	166['85]	184	247['97]	505	366	514['26]
Lisbon	180	240	224['64]	n.d.	242['78]	301	356	435	486	594
Liverpool	80	376	444	493	553	630	704	753	805	856
Lodz	---	16	30	34	57	113	315['97]	408	452	605
London	1,117	2,685	3,227	3,890	4,770	5,638	6,586	7,256	7,488	8,216
Madrid	160	281	271['57]	332	398	470['87]	540	600	751	834
Manchester	90	303	339	351	462	505	645	714	736	766
Marseille	111	194	261	313['72]	360	404	491	551	586	610
Milan	135	242	242	262	322	n.d.	493	579	836	992
Moscow	250	365	352['63]	612	748['82]	799['88]	989['97]	1,533	1,050	2,029['26]
Munich	40	110	148	169	230	349	500	596	631['19]	735['33]
Naples	427['96]	449	417	449	494	n.d.	564	723	722	839
Odessa	6	90	119['63]	121['67]	194['77]	314	405['97]	506	428	421['26]
Paris	581	1,053	1,696	1,852['72]	2,269	2,448	2,714	2,888	2,907	2,891
Prague	75	118	143['57]	157	162	184	202	224['17]	677[2]	849
Riga	30	70	77['63]	102['67]	160	196['88]	256['97]	331	285['23]	378
Rome	163['90]	175	184	244	300	n.d.	463	542	692	1,008
Rotterdam	53	90	106	116	153	202	319	427	511	582
St Petersburg	336['11]	485	539['63]	667['69]	877	1,003['89]	1,267['97]	1,962	722	1,690['26]
Stockholm	76	93	112	136	169	246	301	342	419	502
Vienna	247	444	476['57]	834	1,104	1,365	1,675	2,031	1,866['23]	1,874['34]
Warsaw	100	160	163['63]	252	339	454	638['97]	872	931	1,179
Zürich	12	17	20	57	79	94	151	191	207	250

Fig. 2.2 Population (in thousands) of selected major European cities 1800–1930. ['69], [etc.] Year of data collection different to that indicated in column. [1] Greater Brussels from 1850/51.
Source Palgrave Macmillan (2013), Chandler (1987) for Istanbul 1800–1870

end of the century, as births began to outnumber deaths, cities were able to produce and sustain an increasing proportion of the workforce. In many cities, particularly in Britain—which experienced the highest population growth rate in Europe during this period—migration occurred alongside the natural growth of the settled population.

The scale of rural–urban migration varied significantly from region to region and from city to city. This reflected the uneven process of industrialization as well as the characteristics of local economies and their particular labour demands. Moreover, urban migration was not necessarily the only or preferred option. For example, the persistence of a propertied-owning peasantry in many parts of France meant less pressure to move to the cities, while in southern Italy the greater inclination was for transatlantic emigration. In Britain, urban industrialization first accelerated in the late eighteenth century in the textile mill towns, which were among the first to be mechanized. Manchester, for example, grew from a small market town in the 1770s to become Europe's preeminent industrial metropolis by the 1850s, with the majority of its population coming from the surrounding county of Lancashire and a significant minority from Ireland (Briggs, 1968: 88–138). Similar migration-fuelled growth would later occur in northern and western Europe from the mid-nineteenth century, notably in the Ruhr area, followed by cities in central-eastern Europe in the late nineteenth century, such as Łódź, the 'Polish Manchester' (Young & Kaczmarek, 2008: 58–59), while Moscow and St Petersburg would not experience a significant increase in immigration rates until the turn of the century, following the Russian Empire's belated push for industrial development, the lifting of restrictions on internal movement, and a demographic explosion in the Russian heartland. By the early Soviet era, urban migration would reach levels unparalleled anywhere in the world (Hoffmann, 1994).

Despite the many regional differences, the bulk of urban migration consisted of short-distance, internal movements. This was one of the general conclusions of the German geographer Ernst Georg Ravenstein, whose analysis of population change in the 1881 British census is widely regarded as the first systematic study of migration (Ravenstein, 1885). Having himself emigrated to London in 1852, Ravenstein observed that many migrants moved in stages from village to town to city. He also noted that women constituted the majority of short-distance movers and that people were more likely to undertake longer journeys to larger urban centres offering greater economic opportunities. He later extended his 'laws of migration' to mobility in Europe and North America, whereupon he reaffirmed the prevalence of step-by-step migration, consisting largely of short-distance movements (excluding the transatlantic crossing) and that, of all possible reasons, the main motivation for migration was material improvement (Ravenstein, 1889). 'Migration,' Ravenstein declared, 'means life and progress; a sedentary population stagnation' (ibid: 288).

The reality, however, was more complex than the scenario constructed from static census data. Urban migration was rarely an incremental series of events culminating in a final destination. As research on nineteenth-century migration statistics for German cities has shown, people often moved frequently and rarely in a single direction. Indeed, until 1900, low-skilled workers in some regions, such as the heavily industrialized Ruhr area, continued to work in the countryside during the harvest season to supplement their income and weather economic downturns (Hochstadt, 1999). Meanwhile, many urban jobs, such as those in construction, were seasonal, meaning that large numbers of city dwellers were not permanent residents, but instead engaged in circular migration to other places (Moch, 2003: 128). As in the early modern period, single young women were more likely to migrate to the city, but they made fewer subsequent moves than their male counterparts, who often moved from place to place in search of work. The type of city and economy determined the gender balance of migration. For example, the German industrial city of Duisburg was dominated by male migrants who worked in the steel factories, while Cologne, a historic commercial and administrative centre only 60 kms away, received a greater number of women who were employed as office clerks and domestic servants in middle-class households.

Although numerically less significant overall, there were some important international labour flows. Italians moved to French cities to find work in construction (Milza, 2004) or, in the case of Marseille, on the docks, where they made up almost a quarter of the population (Témime, 1985). Poles, Italians and Dutch found employment in the textile factories and mines of the Ruhr (Klessman, 1986), while the Irish (technically 'internal' migrants at the time) formed a significant part of the workforce in British cities, particularly Liverpool and Manchester, where they occupied some of the least skilled and lowest paid jobs (Dennis, 1986: 35–40). The extent of international mobility varied according to urban and national contexts. French cities, for example, experienced a greater influx of foreign workers than elsewhere, due to the enduring ties of French peasants to the land, chronic labour shortages in key sectors such as mining, and industrialists' desire for a divided and malleable workforce (Noiriel, 2019). International labour migration did not necessarily have to be long distance. This was the case in Roubaix in northern France, where Belgians dominated the workforce in the local textile industry. By 1900, around half of the town's 125,000 inhabitants were Belgian-born or of Belgian descent, while many others commuted across the border daily, taking advantage of the high wages in Roubaix and the lower cost of living in Belgium (Moch, 2003: 132–136).

International labour migration to cities also involved people from outside Europe. Although their numbers were modest compared to those moving within the continent, these migrants played an important role in the expansion of the European economy. With the growth of global maritime trade, Asian and African seafarers became a core part of the workforce in European merchant shipping. They accounted for up to a quarter of a ship's crew, yet earned only a third of their European counterparts' wages despite performing the most arduous tasks (Tabili, 2007). From the mid-nineteenth century onwards, many came ashore in Atlantic and Mediterranean ports in search of better work, albeit in the worst-paid sectors such as tramp shipping. This was the case of the few thousand colonial labourers or 'lascars' from India, Malaya, the Caribbean, Yemen and Somalia who settled in the dock areas of London, Cardiff, Liverpool and South Shields, where they established Britain's earliest Muslim and, through intermarriage with local women, mixed-race communities (Bunnell, 2016; Tabili, 2011). The global dimension of urban migration in nineteenth-century Europe sheds light on a number of issues—such as the racial division of labour and everyday race relations—that are more readily associated with the post-1945 period. Long regarded as statistically insignificant or a mere curiosity of local history, such episodes have received increasing attention from scholars and community heritage projects in recent years, precisely because they reveal the complex and deeper roots of global mobilities and force us to rethink the historical identities of cities (Sect. 5.4.2).

It was in the city that nineteenth-century European societies also came to terms with their own ethnic, religious and linguistic differences. Mass migration brought with it the growth and spread of multi-ethnic and polyglot cities, especially in central-eastern Europe, where population movements and territorial changes in previous centuries had created a complicated demographic patchwork. Here, the short-distance journeys that Ravenstein highlighted in Britain not only contributed to population growth and the formation of new class structures, but also altered the cultural composition of urban society. In the case of the Austro-Hungarian Empire, this could mean a Hungarian professional moving to Vienna or a German-speaking worker moving to Prague. The second half of the nineteenth century also saw a rapid increase in the migration of Eastern European Jews from villages and shtetls to larger cities, reflecting the growing secularization of Jewish life, a shift from traditional crafts and services to industrial and commercial occupations, and, in the case of the Russian Empire, the lifting of restrictions on urban residence (Zipperstein, 1985). In 1897, a fifth of the one million Jews in Russian Poland lived in a single city, Warsaw (ibid: 17–18).

Some of these new cosmopolitan metropolises were simultaneously transformed into major industrial cities, such as Łódź, with its mix of German, Jewish, Polish and Russian entrepreneurs and textile workers. Other cities became regional centres of culture and learning. The borderland city of Chernivtsi, for example, grew considerably in the late nineteenth century thanks to the arrival of Jews, Romanians, Germans, Ukrainians and Poles from the surrounding region, and emerged as a focal point of the Yiddish language movement as well as a nerve centre of Romanian and Ukrainian nationalism (Frunchak, 2010). These rich cultural mosaics flourished against a backdrop of intense urbanization, but often stood in sharp contrast to the rising ethnonationalist sentiments of the time. German journalists and elites from outside Berlin regularly pilloried the city as 'too new', 'too foreign' and 'too Jewish' to be considered a worthy imperial capital (Richie, 1998). Such controversies played out within cities themselves. Fin-de-siècle Vienna was famous for its Jewish intellectuals, most of whom were first- or second-generation immigrants, such as the composer Gustav Mahler, who grew up in the Moravian town of Jihlava, and Sigmund Freud, who was born in Příbor, also in Moravia, to parents from Galicia in what is now Ukraine. Even Viennese gentiles such as the painters Gustav Klimt and Egon Schiele were of immigrant stock: Klimt's father was from Bohemia, while Schiele's father was half-German and his mother of Czech descent. But Vienna was also a stronghold of authoritarian nationalist politics, epitomized by the figure of Karl Lueger, the city's anti-Semitic populist mayor from 1897 to 1910, who campaigned against the immigration of Russian Jews and whose Christian Social Party is considered a precursor of Austro-fascism (Schorske, 1981).

Urban cosmopolitanism was constructed through everyday intergroup relations, both among workers and among cultural elites. But it was also at times a pretext for social grievances and conflict, especially during economic downturns, when minorities risked being scapegoated for majority discontent. Single events, such as the assassination of a national leader, could also trigger waves of ethnically targeted violence. Pogroms in Warsaw, Kiev, Odessa and Kishinev in the late nineteenth and early twentieth centuries, often instigated by the authorities, forced thousands of Russian Jews to leave for Paris, London, New York and Buenos Aires, among other places. In London, the concentration of poor Eastern European Jews in the city's East End—historically an area of foreign settlement with high levels of deprivation and crime—in turn fuelled a xenophobic campaign by sections of the British media and the Conservative Party, culminating in the 1905 Aliens Act, which introduced the country's first comprehensive immigration restrictions. The structural issues of migration and urban poverty were now collapsed into the figure of the foreign immigrant, and powerful tropes such as 'invasion' and 'siege' were increasingly deployed for political gain (Cohen, 1985).

In extreme cases, national governments and international treaties opted for the organized expulsion and replacement of urban populations to resolve the diversity dilemmas of cities. The 1919 race riots in British port cities, triggered by increased competition in the merchant shipping industry after the First World War, led to the repatriation of over 600 colonial workers from Liverpool alone (May & Cohen, 1974: 120). Most notoriously, the population exchange between Turkey and Greece ratified by the Treaty of Lausanne of 1923, would permanently alter the ethnic composition of Thessaloniki, which in the space of a decade went from being the largest Jewish majority urban centre in Europe, with a substantial Turkish Muslim minority, to a Greek majority city (Clark, 2006; Mazower, 2004).[3] The internationally sanctioned ethnic cleansing of Thessaloniki, like that of Izmir across the Aegean, illustrated the instrumental role of cities in early twentieth-century nation-building, but it was also a sign of things to come. With the annihilation of the city's remaining Jewish community by the Nazis in 1943, cosmopolitan Salonika would become mono-ethnic Thessaloniki.

2.5 Colonial Conquest and Mass Emigration: Europeans in Non-European Cities

Until the early nineteenth century, the forced transportation of slaves from Africa to the Americas far exceeded European migration to colonies around the world.[4] Nevertheless, from the sixteenth century onwards, Europeans began to establish permanent settlements in their conquered territories. Contrary to many foundational myths, these urban centres were almost never built on empty terrain but rather on land that was already occupied and used by indigenous peoples. Colonial urban development was an invariably violent and deadly process. Europeans introduced new diseases, land markets and military might, leading to the devastation, dispossession and displacement of local populations. This would leave an enduring legacy on the spatial and racial hierarchies of settler-colonial cities (Edmonds, 2010; Hugill, 2017).

During the mid-nineteenth century, the demise of chattel slavery, the rapid urbanization and industrialization of European settler-colonial societies, and reductions

[3] Many Jews had already opted to emigrate to Palestine and the United States following the Great Fire of Thessaloniki in 1917 which destroyed most of the Jewish neighbourhoods in the city centre.

[4] In 1983, the historian David Eltis estimated that more than eight million Africans were transported to the Americas up to 1820, whereas fewer than two million Europeans had crossed the Atlantic Ocean by this same date (Eltis, 1983: 252–255). Eltis later revised his estimate of the total number of transatlantic slaves upwards to over 11 million, bringing the end date forward to 1867 and including the approximately 1.5 million people who died en route (Eltis, 2001).

in the time and cost of sea travel led to a significant increase in international labour migration. Huge numbers of Europeans headed for rapidly expanding cities, such as New York and Chicago in the United States, Toronto and Montreal in Canada, Buenos Aires in Argentina, São Paulo in Brazil, Johannesburg in South Africa and Sydney and Melbourne in Australia. Between 1820 and 1930, it is estimated that 55–60 million Europeans crossed the Atlantic alone, about a quarter of whom later returned home (Bade, 2003: 97–98; Moch, 2003: 156). The largest groups were Irish and Germans in the mid-nineteenth century, and Poles and Italians from the 1870s onwards.

Chicago is seen as a paradigmatic case of how immigrant men and women made a fundamental contribution to urban growth and industrial development in the United States. From its origins as a tiny trading post between First Nation Americans and European settlers, until the former were forcibly relocated by the US government in the mid-1830s (Cronon, 1992: 23–30), Chicago developed rapidly in the space of a few decades to become the main commercial and manufacturing centre of the Midwest. By 1910, it had become the second largest city in the country with a population of over two million. At that time, 77 per cent of Chicago's population were first- or second-generation immigrants of various European nationalities, including (in descending order) Germans, Italians, Irish, Poles and Swedes (Harzig, 1997). The emergence of the multinational city, complete with its 'ethnic' neighbourhoods and related issues such as 'assimilation', made international immigration a foundational focus of Chicago School urban sociology (Thomas & Znaniecki, 1996).

Not all transcontinental migration involved crossing the Atlantic or heading for the southern hemisphere. Russian voluntary and involuntary migration to Siberia in the nineteenth century fuelled the growth of cities such as Omsk, Irkutsk and Vladivostok, all of which had populations of over 100,000 by 1917 (Naumov, 2006: 99). It is also often forgotten that significant numbers of southern Europeans had already settled in North African cities prior to the colonization of the region by the French, Spanish, Italians and British. Here they found employment and commercial opportunities, as well as a favourable climate, similar food and greater freedoms than at home (Triulzi, 1971: 165–66). Italians in early nineteenth-century Tunis, for example, were an extremely heterogenous group. Making up around a third of the 8,000-strong European population in the 1830s, the group included freed slaves, political exiles, skilled and unskilled labourers, court functionaries and a sizeable Livornese Jewish community that had dominated local commerce since the seventeenth century (ibid). In an ironic twist of fate, there was even a 'boat people crisis' in 1838, when a ship loaded with Maltese (among the least skilled and most unwanted migrants in North Africa) was forced to remain in rough seas until the Tunisian authorities, under pressure from the British, finally allowed the vessel to enter the port of Tunis (Clancy-Smith, 2011).

Following colonial conquest, labour migration from Southern Europe to North African cities continued, but was now accompanied by elite groups of administrators, military personnel, engineers and industrialists. Nineteenth-century colonial projects throughout Africa and Asia involved the construction of command centres, either in the form of new capital cities or modern districts attached to existing settlements

(King, 1985). As well as operating as the centres of administrative, economic and military coordination, these 'European cities' were designed to house the ruling class and demonstrate the technological and planning superiority of the colonial powers. As Frantz Fanon memorably wrote of Algiers in *The wretched of the world* (Fanon, 1963), the new quarters were also a physical testament to colonial violence, crystallizing the racial segregation between European and indigenous residents and regulating the movement of different groups within the city.

Rabat in Morocco is a particular case in point. After the city was chosen as the new capital of the French protectorate in 1912, the colonial authorities hastily built a new administrative and residential district adjacent to the medina, the original settlement. This allowed for the arrival of thousands of colonial migrants from France, who by 1931 accounted for over 40% of the city's 55,000 inhabitants (Brown, 1976: 47). The transformation of Rabat resulted in the juxtaposition of two different urban worlds. On the one hand, an exemplary European quarter of elegant whitewashed buildings and wide avenues showcased the engineering skills, aesthetic sensibilities and sanitary concerns of the French colonial state; on the other, the indigenous medina, with its cramped houses and narrow streets, was preserved as a testimony to the city's architectural heritage and as an attraction for Western tourists. This distinction reflected a deep disparity in living conditions: healthy, modern facilities for the European migrants; unhealthy and overcrowded accommodation for the Moroccans (Abu-Lughod, 1980; Rabinow, 1989).

In short, the transformation of Rabat during the French protectorate offers us a very different perspective on the relationship between migration and cities. Here, the dominant figure was not the European migrant worker, as in Chicago, but the member of an expatriate colonial elite or at least someone who benefited from the colonial social order. The principal goal of colonialism in Rabat was not to build a permanent settler society from scratch but rather 'to ensure the uninterrupted persistence of metropolitan accumulation' (Hugill, 2017: 4). Recent research has sought to complicate the Manichean portrayal of colonial cities, in particular by emphasizing the national and class differences among European residents. Ilaria Giglioli, for example, draws attention to the ambivalence of colonial discourse regarding working-class Sicilian migrants in late nineteenth-century Tunis. On the one hand, this group was depicted as being culturally similar to the natives, on the other hand, they were considered Christian allies who could potentially bolster French interests locally (Giglioli, 2017).

The movement of European settlers to colonial cities in Africa and Asia reached its peak in the early twentieth century. The Second World War would mark the end of this migration, as anti-colonial independence movements gathered strength around the world. The subsequent process of decolonization would see most white European urbanites in North Africa return to the metropoles, often fleeing violent liberation struggles such as the Algerian War of Independence, and sometimes forcibly repatriated, as in the case of the 20,000 Italians expelled from Libya by Muammar Gaddafi in 1970. Many more would leave voluntarily, usually at odds with the new political order. The parallel movement of former colonial subjects to European cities would instead become a key feature of international migration in the second half of the twentieth century.

2.6 International and Internal Migrations and Urban Change After 1945

The Second World War marks a turning point in the history of urban migration in Europe. Not only did the war destroy the urban fabric and economies and decimate the populations of many cities, it also erased the traces of centuries of mobility and cultural diversity. Following the Holocaust and the displacement of minority (and sometimes majority) groups during and immediately after the war, cities across Europe—from Amsterdam to Warsaw, Berlin to Prague, Kaliningrad to Rijeka—that had previously been characterized by their linguistic, religious and ethnic mix became increasingly homogeneous. The main centres and translocal networks of European Jewry disappeared forever. As a combined result of the brutal German-Romanian occupation and deportations to extermination camps, the Jewish population of Odessa, previously one of the largest in Europe, was reduced to a few thousand by the end of the war (King, 2011).

The immediate aftermath of the conflict saw the mass transfer of people to and from cities, many of them uprooted by territorial readjustments. Istrian Italians moved in large numbers to Trieste, Venice and especially Rome, where they formed the nucleus of the new Giuliano-Dalmata district (Fidanzia, 2003). The German population of Breslau, now called Wrocław and under Polish administration, was expelled westwards and replaced by Poles from Lvov (formerly Lwow, now Lviv), which had become part of the Soviet Union (Thum, 2011).

If cities were places of refuge for some, they were also instrumental in reorganizing the populations and economies of shattered nations. In 1946, the British government began recruiting Eastern European women and men from displaced persons camps in Germany and Austria to work in hospitals, domestic service and the textile industry in cities such as Bradford, Coventry and Manchester, in an attempt to address the country's severe labour shortages (McDowell, 2005). Not for the first time in history, the distinction between 'refugee' and 'economic migrant' was blurred as 'European Voluntary Workers' entered Britain's urban workforce alongside a new wave of Irish immigrants (Kay & Miles, 1988).

Although these initial flows contributed to the post-war recovery, from the mid-1950s they were overshadowed by the massive city-bound migrations that enabled the subsequent expansion of European economies. This continent-wide movement can be divided into three broad types: first, migrations from European colonies or former colonies, mainly to Britain, France and the Netherlands; second, recruitment schemes that brought workers from Mediterranean countries to Germany, Switzerland, Austria, the Benelux countries and Scandinavia; and third, long-distance internal migrations in southern and eastern Europe. As in the past, urban newcomers, particularly in the north-western quadrant of Europe, tended to fill low-skilled jobs vacated by local workers who had moved into white-collar jobs and sometimes out of the city itself.

The bulk of international labour migration to Britain, France and the Netherlands came from the collapsing colonial empires of these countries. Around 180,000 Surinamese settled in the Dutch Randstad in the 1970s (Moch, 2003: 178), while by 1982 there were 800,000 Algerians in France, mainly concentrated in the Paris region and Marseille, but also in industrial towns such as Roubaix (Noiriel, 2019: 145). The arrival of the former German troopship Empire Windrush at Tilbury Docks near London in 1948, with 802 Caribbean passengers on board, a quarter of whom were women, is generally regarded as a seminal moment in the history of black migration to Britain. Beyond its symbolic significance (for a critical discussion, see Hammond Perry, 2015), the ship's arrival was an exquisitely *urban* event: many of the passengers were temporarily housed in underground air-raid shelters in south London and later remained in the Brixton area where they registered at the local labour exchange. Over the next two and a half decades, hundreds of thousands of women and men from the former British Empire settled in towns and cities across the country. As citizens of 'the United Kingdom and the Colonies', all had a legal right of entry, although this was restricted in the 1960s and finally removed by the Immigration Act of 1971. Some were recruited to work in factories, public transport or the newly established National Health Service, but many followed in the footsteps of family and friends, lured by stories of success or readily available jobs.[5] Certain patterns emerged: Indians were most numerous in the north-west Midlands, London and West Yorkshire, many Pakistanis concentrated in northern mill towns such as Bradford and Oldham, while East African Asians settled mainly in Leicester and north-west London (Herbert, 2008).

Guest worker schemes similarly altered the demographic composition of many cities, enabling otherwise unimaginable rates of economic growth (Berger & Mohr, 1975; Castles, 1986). With its rapidly expanding economy and numerous bilateral agreements with sending countries, West Germany had by far the most organized labour recruitment system in Europe, which between 1955 and 1973 brought some 2.6 million workers from Mediterranean Europe, Turkey and North Africa to work in factories and public works. Initially, the state, employers and workers expected this migration to be short-lived: the *Gastarbeiter* would stay for two years, acquire income and skills, and then leave. However, it soon became clear that it was too expensive to train replacements and many workers simply did not return home, so in 1965 the German government granted permanent residence to foreign migrants and a few years later provided for family reunification. Of the 750,000 guest workers

[5] According to the 1971 UK census, 237,000 residents were born in the Caribbean, 322,000 in India, 139,900 in Pakistan, 59,500 in Kenya (the majority of whom were of South Asian origin), and 104,700 in other parts of Africa (excluding South Africa). It is worth noting that together these groups made up 28.9 per cent of the total foreign-born population at the time (and 1.6 per cent of the UK population), and were far outnumbered by people born in Europe (including the Republic of Ireland), who together made up 50.9 per cent of the foreign-born population (Rendall & Salt, 2005: 134).

who came to Germany from Turkey in the 1960s and early 1970s, about half stayed, eventually settling in less desirable urban areas such as around central railway stations or in outer suburbs (Thomsen Vierra, 2018). In 2001, more than 25 years after the end of the recruitment programme, the vast majority of Germany's 1.9 million Turkish citizens—a figure inflated by subsequent arrivals of family members and political refugees, including many Kurds—continued to live in the cities and towns where the first guest workers had settled (Yurdakul, 2009).

The third major flow towards cities—internal migration—affected the entire continent in the post-war period,[6] but was particularly pronounced in southern and eastern Europe, where foreign immigration was minimal. In the Soviet Union, urban migration was theoretically restricted by the internal *propiska* passport system, housing shortages and 'closed cities', which included Moscow and Leningrad (Gang & Stuart, 1999). In reality, pressure to leave rural areas and the desire for city life, on the one hand, and high labour demand and the emergence of a shadow 'second' economy, on the other, coupled with ways of circumventing restrictions, from corruption to marriages of convenience, meant that internal migration still accounted for over half of urban growth between 1959 and 1979 (Buckley, 1995; Sahadeo, 2019; Siegelbaum & Moch, 2014). Meanwhile, in Spain, Greece and Italy, internal migrants now outnumbered emigrants as millions of rural and smalltown dwellers moved to the main urban and industrial areas of their respective countries. Massive movements of people left Andalusia, Extremadura and Galicia for Madrid, Barcelona and the Basque Country, the Greek mainland and islands for Greater Athens, the Italian South and North East for the industrial triangle of Milan, Turin and Genoa and other major cities with expanding public sectors such as Rome. There were often major differences between the sending regions and the destination cities in terms of history, culture and language, and the distances travelled tended to be much greater than in the past, sometimes exceeding 1,000 kms. In the case of Milan, at the heart of Italy's post-war 'economic miracle', newcomers quickly found casual work on construction sites or in service industries, before moving into more secure positions in the rapidly expanding manufacturing sector. In response to a severe housing shortage and under pressure from burgeoning labour and social movements that often saw southern workers and their families at the forefront of struggles, the state channelled funds into the construction of vast housing estates for some 300,000 new inhabitants on the outskirts of the city (Foot, 2001). As a result of internal migration, the urban landscape of Milan, like that of other northern Mediterranean cities, was permanently transformed. Nothing of the kind was built for Commonwealth citizens in Birmingham, for guest workers in Munich, or for the foreign immigrants who began arriving in large numbers in Milan in the 1980s.

The historical relationship between urban migration and demographic growth changed considerably in the post-war period. Overall, the continent's urban population appeared to grow steadily. In 1900, there were only seven European cities with more than one million inhabitants: by 1950, 16 cities had crossed this threshold, and

[6] For example, the vast majority of migrants arriving in Paris between 1945 and 1960 came from other French regions, with one in six coming from Brittany (Moch, 2012).

by 1970 the number had reached 26 (Palgrave Macmillan, 2013). Some of the largest increases were in southern and eastern Europe. Mediterranean industrial centres such as Milan and Barcelona grew by over 35 per cent during the two decades of mass internal migration, while cities in Yugoslavia and the Eastern Bloc such as Belgrade, Bucharest, Sofia and Kiev more than doubled their populations between 1950 and 1980. Elsewhere, however, and particularly in the UK, many cities experienced demographic stagnation or decline due to counter-urbanization and falling fertility rates.[7]

While *internal* migration continued to drive population growth, the impact of *international* migration was negligible. For example, the main boroughs of immigrant settlement in London in the 1960s, such as Lambeth, Tower Hamlets and Hackney, had already experienced significant population decline since the interwar period as residents moved to suburbs and new towns in the Home Counties or emigrated abroad. The population chart for this period (Fig. 2.3) should be interpreted with caution because, unless stated otherwise, the figures refer to the city within its administrative boundaries. Just as the city had outgrown its walls in the nineteenth century, it was now no longer confined to its municipal borders: industrial and residential sprawl had led to the urbanization of the surrounding region. In short, before middle-class revanchism and inner-city redevelopment gathered pace in the 1980s, international migration effectively prevented further demographic decline in Europe's urban cores.

Whatever their impact on urban growth, it was to the city that these three migration streams were directed and, as we shall see in the following chapters, it was in the city that related issues such as employment, housing and cultural diversity unfolded. All migrant groups, whether they moved to London or Leningrad, Frankfurt or Turin, were subject to varying forms and degrees of discrimination and prejudice. The subsequent conflicts and acts of resistance—from riots to labour disputes, from antiracist organizations to street festivals—were again fundamentally urban phenomena.

These migration histories have also provided important alternative frameworks to conventional nation-centred accounts of post-war societal change. Events and places such as the arrival of the Empire Windrush in 1948 or Milan's Central Station (the disembarkation point for hundreds of thousands of southern Italians) have been etched into collective urban memory and are often mobilized in cultural and political contexts to counter nativist accounts of the past and present. But as historical narratives, the three major post-war migrations have their limits. In particular, they run the risk of affirming the idea that cities were affected by a single type of migration, and that employment and settlement trajectories could somehow be explained in terms of migrants' regional or ethnic origins.

Recent historical research has complicated our understanding of post-war urban migration by highlighting the internal diversity of migrant groups alongside intersectional dimensions of gender, class and age. It has also revealed the multiple

[7] A similar trend of population decline would occur in southern European cities after the 1970s, following the demise of mass internal migration and compounded by some of the lowest birth rates in the world. All major cities in northern and central Italy, with the exception of Rome, lost around a quarter of their population between 1971 and 2011.

	1940/1	1950/1	1960/1	1970/1	1980/1	1990/1	2000/1	2009/10
Amsterdam	794	804	865	820	717	713	739['03]	767
Athens	481	565	628	867	886	772	745	721['07]
Barcelona	1,081	1,280	1,558	1,745	1,755	1,668	1,583['03]	1,616['08]
Belgrade	n.d.	368	585	746	1,088	1,136	1,687	1,631
Berlin	4,332	3,337[1]	3,261[1]	3,208[1]	3,057[1]	3,438	3,392['02]	3,443
Birmingham	1,053	1,113	1,107	1,015	1,007	994	990['02]	1,037
Brussels[2]	913['38]	956	1,020	1,075	1,000	954	998['03]	1,441['07]
Bucharest	648	886	1,226	1,475	1,929	1,934	1,928['04]	1,944
Budapest	1,163	1,571	1,805	1,945	2,059	2,017	1,708['03]	1,717
Dortmund	537	507	648	640	609	600	590['99]	580
Frankfurt	547	532	692	670	629	647	645['99]	680
Glasgow	1,132	1,090	1,055	898	766	687	578	593
Hamburg	1,682	1,606	1,832	1,794	1,649	1,661	1,728['02]	1,774
Istanbul	794	983	1,467	2,137	2,773	6,220	9,556['03]	10,823['07]
Kiev	846	n.d.	1,104['59]	1,632	2,248	2,643	2,639['04]	2,724
Lisbon	702	790	817	760	807	678	565	485
London	8,700['38]	8,348	8,172	7,452	7,678	6,803	7,355['02]	7,825
Madrid	1,089	1,618	2,260	3,146	3,188	2,991	3,093['03]	3,213['08]
Manchester	728	703	661	544	449	433	393	499
Marseille	620['36]	661['54]	778['62]	889['68]	879['82]	800	807['99]	839['06]
Milan	1,116['36]	1,260	1,583	1,724	1,635	1,549	1,247['02]	1,302
Moscow	4,137	n.d.	5,046['59]	7,061	8,203	8,747	10,101['02]	10,536
Naples	866['36]	1,011	1,183	1,233	1,211	1,208	1,008['02]	963
Odessa	604	n.d.	667['59]	892	1,072	1,096	1,029	991
Paris	2,830['36]	2,850	2,790	2,489	2,189	2,152	2,148['99]	2,181['06]
Prague	928['35]	922['47]	1,005	1,080	1,193	1,212	1,162['02]	1,249
Rome	1,156['36]	1,652	2,188	2,800	2,831	2,828	2,460	2,734
St. Petersburg	3,191	n.d.	3,321['59]	3,950	4,676	4,437	4,660['99]	4,591
Stockholm	591	744	807	740	647	679	762['03]	789['07]
Vienna	1,918	1,616	1,628	1,620	1,531	1,540	1,608	1,687
Warsaw	1,266	601	1,136	1,308	1,596	1,654	1,688['02]	1,711

Fig. 2.3 Population (in thousands) of selected major European cities 1940–2010.
[1] East and West Berlin combined. [2] Greater Brussels throughout.
Source Palgrave Macmillan (2013)

journeys, strategies and translocal ties of individual migrants, and the different types of migration that affected individual cities (De Bock, 2018; Miller, 2018). Compared to earlier periods, urban historians have a much richer range of sources at their disposal, including oral histories, migrant literatures, and empirical research conducted at the time. In her study of migration to Ghent in the 1960s and 1970s, Jozefien De Bock (2015) draws on administrative files and her own in-depth interviews with migrants from various Mediterranean countries to unpack the master category of 'guest worker'. Contrary to popular assumptions, only a small minority were recruited by local employers, while the rest had diverse migration experiences. Most had arrived as tourists, others as students and almost all had overstayed and later regularized their status, often after a long period of 'illegality'. Some were migrant pioneers, others had followed, some had come directly, others had moved through different countries and cities, some had settled, but many had left within a year of arrival.

Urban migration has changed significantly since the 1970s. The global recession after 1973 spelt the end of post-war economic growth and coincided with the end of state recruitment programmes and a marked decline in internal migration.

The decade also saw increasing restrictions on immigration from former European colonies. Five key shifts can be identified. First, the 1980s saw the increasing globalization and diversification of labour flows across much of Europe, as migrants no longer followed routes established by international bilateral agreements or colonial histories. Second, new destinations for foreign immigration emerged as Italy, Greece, Spain and Portugal transitioned from labour-exporting to labour-importing countries. As before, migrants were heavily concentrated in cities, but with the restructuring of urban economies, they were now increasingly employed in services and light industry. Third, the collapse of state socialism after 1989 led to an increase in international labour migration from Central and Eastern Europe, which accelerated with the accession of former Eastern Bloc countries to the European Union after 2004. Fourth, the tightening of immigration controls at national and EU level, coupled with increased freedom of movement within the European Union itself, has created new forms of intra-urban mobility, such as the Erasmus university exchange programme, but also immobility for those non-EU citizens excluded from the benefits of open borders. Finally, following a global increase in asylum seekers and refugees since the 1990s, European cities have seen the settlement of people from countries such as the former Yugoslavia, Iraq, Afghanistan, Syria, Eritrea and Somalia, who have often become the most visible and publicly scrutinized section of the urban migrant population, especially in the wake of Europe's 'refugee crisis' of the 2010s.

The remainder of this book explores the multidimensional relationship between these recent processes of mobility and Europe's cities. The chapters that follow will also necessarily take into account the radical changes to urban regions over the last four decades, which, as Glick Schiller and Çağlar (2011) remind us, migrants themselves have played an active role in shaping. In his comparative study of internal migration to Milan in the 1960s and foreign immigration to the same city in the 1990s, historian John Foot describes how both groups of migrants were 'wanted but not welcome.' However, he acknowledges that this slogan alone does little to capture the differences and similarities in the experiences of the city's southern Italian and foreign-born residents: 'It is impossible to deal with the complications of these two immigrations without grappling with the changes that have transformed Milan itself' (Foot, 2001: 70).

2.7 Conclusion

The history of migration in Europe cannot be separated from the history of its cities. This chapter has focused primarily on the different types of migration *to* and *from* cities over the last five centuries. It has considered the settlement of migrants *in* cities and their changing influence *on* urbanization and change. It has highlighted the diversity between and within urban migration flows, especially during periods such as the nineteenth century and the post-war decades, which are often sweepingly associated with epic mass movements, and it has pointed to some of the networks that existed between cities long before our current global era. It has also drawn attention to the turbulent dimension of urban migration throughout history. While urban and national governments have both attempted to channel migration to cities at various times in the past, either by offering sweeteners such as urban citizenship or by recruiting labour, their capacity to regulate migration has been more limited, as exemplified by the incessant migration towards the 'closed cities' of Moscow and Leningrad during the Soviet era. Moreover, the consequences of 'managed' flows were difficult to predict, as evidenced by the failure of guest worker programmes to monopolize the supply of migrant labour or to anticipate permanent settlement.

Tracing the historical complexity of the relationship between migration and cities raises critical awareness of the deep-seated role of migration in changing the course of urban history, and cautions us against making rash claims about the unique novelty of present-day situations. De Bock, for example, argues that historians should overcome their scepticism and incorporate the concept of 'superdiversity' into their research 'as an analytical tool [that can illuminate] more clearly the interplay of the many different factors that condition immigrants' lives' and one that shifts the emphasis away from the 'methodological nationalism' and 'ethnic groupism' that has structured much of migration scholarship (2015: 584, 586). While this assertion is somewhat overstated—historians have long been attentive to the nuances and variables that structure urban migration—it is certainly the case that, from a historical perspective, the city rather than the nation is better placed to address such issues.

References

Abu-Lughod, J. (1980). *Rabat: Urban apartheid in Morocco*. Princeton University Press.

Bade, K. J. (2003). *Migration in European history*. Blackwell.

Benedict, P. (Ed.). (1989). *Cities and social change in early modern France*. Routledge.

Berger, J., & Mohr, J. (1975). *A seventh man. A book of images and words about the experience of migrant workers in Europe*. Penguin.

Blackburn, R. (1997). *The making of new world slavery: From the Baroque to the modern 1492–1800*. Verso.

Bonazza, G. (2019). *Abolitionism and the persistence of slavery in Italian states, 1750–1850*. Palgrave Macmillan.

Briggs, A. (1968). *Victorian cities*. Penguin.

Brown, K. L. (1976). *People of Salé: Tradition and change in a Moroccan city, 1830–1930*. Manchester University Press.

Buckley, C. (1995). The myth of managed migration: Migration control and market in the Soviet period. *Slavic Review, 54*(4), 896–916.

Bunnell, T. (2016). *From world city to the world in one city: Liverpool through Malay lives*. Wiley.

Castles, S. (1986). The guest-worker in western Europe: An obituary. *The International Migration Review, 20*(4), 761–778.

Chandler, T. (1987). *Four thousand years of urban growth: An historical census*. St David's University Press.

Clancy-Smith, J. A. (2011). *Mediterraneans: North Africa and Europe in an age of migration, c. 1800–1900*. University of California Press.

Clark, B. (2006). *Twice a stranger: The mass expulsions that forged modern Greece and Turkey*. Harvard University Press.

Clark, P. (2009). *European cities and towns 400–2000*. Oxford University Press.

Cohen, S. (1985). Antisemitism, immigration controls and the welfare state. *Critical Social Policy, 5*(13), 73–92.
Cornelissen, C., Kümin, B., & Rospocher, M. (Eds.). (2022). *Migration and the European city: Social and cultural perspectives from early modernity to the present*. de Gruyter.
Cronon, W. (1992). *Nature's metropolis: Chicago and the great west*. WW Norton & Company.
De Bock, J. (2015). Not all the same after all? Superdiversity as lens for the study of past migrations. *Ethnic and Racial Studies, 38*(4), 583–595.
De Bock, J. (2018). *Parallel lives revisited: Mediterranean guest workers and their families at work in the neighbourhood, 1960–1980*. Berghahn.
De Munck, B., & Winter, A. (Eds.). (2012). *Gated communities? Regulating migration in early modern cities*. Ashgate.
Dennis, R. (1986). *English industrial cities of the nineteenth century*. Cambridge University Press.
De Vries, J. (1984). *European urbanization 1500–1800*. Harvard University Press.
Dines, N. (2012). *Tuff city: Urban change and contested space in central Naples*. Berghahn.
Dresser, M. (2001). *Slavery obscured: The social history of the slave trade in an English provincial port*. Bloomsbury.
Du Bois, W. E. B. (1935). *Black reconstruction*. Harcourt, Brace and Company.
Edmonds, P. (2010). *Urbanizing frontiers: Indigenous peoples and settlers in 19th-century Pacific rim cities*. UBC Press.
Eltis, D. (1983). Free and coerced transatlantic migrations: Some comparisons. *The American Historical Review, 88*(2), 251–280.
Eltis, D. (2001). The volume and structure of the transatlantic slave trade: A reassessment. *The William and Mary Quarterly, 58*(1), 17–46.
Fanon, F. (1963). *The wretched of the earth*. Grove Weidenfeld.
Fidanzia, R. (2003). Il quartiere Giuliano-Dalmata di Roma tra memoria e attualità (The Giuliano-Dalmata neighbourhood between memory and the present). *Storia del mondo – Periodico telematico di storia e scienze umane, 3*, 1–7.
Foot, J. (2001). *Milan since the miracle: City, culture and identity*. Berg.
Frunchak, S. (2010). Commemorating the future in post-war Chernivtsi. *East European Politics and Societies, 24*(3), 435–463.
Gabaccia, D. (2015). Time and temporality in migration studies. In C. B. Brettell & J. F. Hollifield (Eds.), *Migration theory: Talking across disciplines* (pp. 37–66). Routledge.
Gang, I. N., & Stuart, R. C. (1999). Mobility where mobility is illegal: Internal migration and city growth in the Soviet Union. *Journal of Population Economics, 12*(1), 117–134.
Giglioli, I. (2017). Producing Sicily as Europe: Migration, colonialism and the making of the Mediterranean border between Italy and Tunisia. *Geopolitics, 22*(2), 407–428.
Gilroy, P. (1993). *The black Atlantic. Modernity and double consciousness*. Verso.
Glick Schiller, N., & Çağlar, A. (2011). *Locating migration: Rescaling cities and migrants*. Cornell University Press.
Goose, N. (1982). The "Dutch" in Colchester: The economic influence of an immigrant community in the sixteenth and seventeenth centuries. *Immigrants & Minorities: Historical Studies in Ethnicity, Migration and Diaspora, 1*(3), 261–280.
Greefs, H., & Winter, A. (Eds.). (2019). *Migration policies and materialities of identification in European cities: Papers and gates, 1500s–1930s*. Routledge.
Hammond Perry, K. (2015). *London is the place for me: Black Britons, citizenship, and the politics of race*. Oxford University Press.
Harzig, C. (Ed.). (1997). *Peasant maids, city women. From the European countryside to urban America*. Cornell University Press.
Henrich, E., & Sampson, J. M. (Eds.). (2019). *History, historians and the immigration debate: Going back to where we came from*. Palgrave Macmillan.
Herbert, J. (2008). *Negotiating boundaries in the city: Migration, ethnicity, and gender in Britain*. Ashgate.

Herlihy, P. (1976). Russian wheat and the port of Livorno, 1794–1865. *Journal of European Economic History, 5*(1), 45–68.

Hochstadt, S. (1999). *Mobility and modernity. Migration in Germany, 1820–1989*. University of Michigan Press.

Hoffmann, D. L. (1994). *Peasant metropolis. Social identities in Moscow, 1929–1941*. Cornell University Press.

Hohenberg, P. M., & Lees, L. H. (1995). *The making of urban Europe, 1000–1994*. Harvard University Press.

Hugill, D. (2017). What is a settler-colonial city? *Geography Compass, 11*(5), 1–11.

Kasaba, R. (2009). *A moveable empire: Ottoman nomads, migrants and refugees*. University of Washington Press.

Kay, D., & Miles, R. (1988). Refugees or migrant workers? The case of the European volunteer workers in Britain (1946–1951). *Journal of Refugee Studies, 1*(3–4), 214–236.

Killingray, D. (2011). African slaves in Britain in the early modern period. In K. J. Bade, P. C. Emmer, L. Lucassen, & J. Oltmer (Eds.), *The encyclopedia of migration and minorities in Europe: From the 17th century to the present* (pp. 211–213). Cambridge University Press.

King, A. (1985). Colonial cities: Global pivots of change. In R. Ross & G. Telkamp (Eds.), *Colonial cities: Essays on urbanism in colonial context* (pp. 7–32). Springer.

King, C. (2011). *Odessa: Genius and death in a city of dreams*. W.W. Norton & Company.

Klessman, C. (1986). Comparative immigrant history: Polish workers in the Ruhr area and the north of France. *Journal of Social History, 20*(2), 335–353.

Kohl, J. G. (1841). *Reisen in Südrußland* (Travels in southern Russia). Arnoldische Buchhandlung.

Kohl, J. G. (1844). *Russia: St Petersburg, Moscow, Kharkoff, Riga, Odessa, The German provinces on the Baltic, the Steppes, the Crimea, and the interior of the empire*. Chapman and Hall.

Linebaugh, P., & Rediker, M. (2000). *The many-headed hydra: Sailors, slaves, commoners, and the hidden history of the revolutionary Atlantic*. Beacon Press.

Longmore, J. (2007). "Cemented by the blood of a negro"? The impact of the slave trade on eighteenth-century Liverpool. In D. Richardson, S. Schwarz, & A. Tibbles (Eds.), *Liverpool and transatlantic slavery* (pp. 227–251). Liverpool University Press.

Lowe, L. (2015). *The intimacies of four continents*. Duke University Press.

Lucassen, J., & Lucassen, L. (Eds.). (1997). *Migration, migration history, history: Old paradigms and new perspectives*. Peter Lang.

Lucassen, J., & Lucassen, L. (2009). The mobility transition revisited, 1500–1900: What the case of Europe can offer to global history. *Journal of Global History, 4*(3), 347–377.

Lucassen, L. (2005). *The immigrant threat: The integration of old and new migrants in Western Europe since 1850*. University of Illinois Press.

McDowell, L. (2005). *Hard labour: The forgotten voices of Latvian migrant "volunteer" workers*. UCL Press.

McKeown, A. (2004). Global migration 1846–1940. *Journal of World History, 15*(2), 155–189.

May, R., & Cohen, R. (1974). The interaction between race and colonialism: A case study of the Liverpool race riots of 1919. *Race and Class, 16*(2), 111–126.

Mazower, M. (2005). *Salonica, city of ghosts: Christians, Muslims and Jews 1430–1950*. Alfred A. Knoff.

Miller, J. (2008). *Urban societies in east-central Europe, 1500–1700*. Ashgate.

Miller, J. A. (2018). *Turkish guest workers in Germany: Hidden lives and contested borders, 1960s to 1980s*. Toronto University Press.

Milza, P. (2004 [1993]). *Voyage en Ritalie* (Trip to Ritalie). Éditions Payot.

Moch, L. P. (2003). *Moving Europeans: Migration in Western Europe since 1650*. Indiana University Press.

Moch, L. P. (2012). *The pariahs of yesterday: Breton migrants in Paris*. Duke University Press.

Morgan, K. (2000). *Slavery, Atlantic trade and the British economy, 1660–1800*. Cambridge University Press.

Naumov, I. G. (2006). *The history of Siberia*. Routledge.

Noiriel, G. (2019). The role of immigration in the making/unmaking of the French working class (nineteenth and twentieth Centuries). In E. Henrich & J. M. Simpson (Eds.), *History, historians and the immigration debate: Going back to where we came from* (pp. 133–151). Palgrave Macmillan.

Macmillan, P. (2013). *International historical statistics*. Palgrave Macmillan.

Prak, M. (2018). *Citizens without nations. Urban citizenship in Europe and the world, c.1000–1789*. Cambridge University Press.

Rabinow, P. (1989). *French modern: Norms and forms of the social environment*. University of Chicago Press.

Ravenstein, E. G. (1885). The laws of migration. *Journal of the Statistical Society of London, 48*(2), 167–227.

Ravenstein, E. G. (1889). The laws of migration. Second paper. *Journal of the Royal Statistical Society, 52*(2), 241–305.

Rendall, M., & Salt, J. (2005). The foreign-born population. In R. Chappell (Ed.), *Focus of people and migration 2005 edition* (pp. 131–151). Palgrave Macmillan.

Richie, A. (1998). *Faust's metropolis: A history of Berlin*. Harper Collins.

Sahadeo, J. (2019). *Voices from the Soviet edge: Southern migrants in Leningrad and Moscow*. Cornell University Press.

Salzberg, R. (2018). Mobility, cohabitation and cultural exchange in the lodging houses of early modern Venice. *Urban History, 46*(3), 398–418.

Schorske, C. E. (1981). *Fin-de-Siècle Vienna: Politics and culture*. Vintage Books.

Siegelbaum, L. H., & Moch, L. P. (2014). *Broad is my native land: Repertoires and regimes of migration in Russia's twentieth century*. Cornell University Press.

Tabili, L. (2007). Lascars and black seamen. In D. Dabydeen, J. Gilmore & C. Jones (Eds.), *The Oxford companion to black British history*. Oxford University Press.

Tabili, L. (2011). *Global migrants, local culture: Natives and newcomers in provincial England, 1841–1939*. Palgrave Macmillan.

Témime, E. (1985). Marseille, ville de migrations (Marseille, city of migrations). *Vingtième Siècle. Revue d'histoire, 7*, 37–50.

Thomas and Znaniecki, 1996 Thomas, W. I., & Znaniecki, F. (1996 [1918–1920]). *The Polish peasant in Europe and America*. University of Illinois Press.

Thomsen Vierra, S. (2018). *Turkish Germans in the Federal Republic of Germany: Immigration, space and belonging, 1961–1990*. Cambridge University Press.

Thum, G. (2011). *Uprooted: How Breslau became Wrocław during the century of expulsions*. Princeton University Press.

Tilly, C. (1978). Migration in modern European history. In W. H. McNeil & R. S. Adams (Eds.), *Human migration* (pp. 48–72). Indiana University Press.

Todorov, N. (1983). *The Balkan city, 1400–1900*. University of Washington Press.

Torpey, J. C. (2000). *The invention of the passport: Surveillance, citizenship and the state*. Cambridge University Press.

Triulzi, A. (1971). Italian-speaking communities in early nineteenth century Tunis. *Revue de l'Occident musulman et de la Méditerranée, 9*, 153–184.

van Lottum, L. (2007). *Across the North Sea: The impact of the Dutch Republic on international labour migration*. Aksant.

Young, C., & Kaczmarek, S. (2008). The socialist past and postsocialist urban identity in central and eastern Europe. The case of Łódź, Poland. *European Urban and Regional Studies, 15*(1), 53–70.

Yurdakul, G. (2009). *From guest workers into Muslims: The transformation of Turkish immigrant associations in Germany*. Cambridge Scholars.

Zelinsky, W. (1971). The hypothesis of the mobility transition. *Geographical Review, 61*(2), 219–249.

Zipperstein, S. J. (1985). *The Jews of Odessa: A cultural history 1794–1881*. Stanford University Press.

Chapter 3
Migration and the Transformation of Urban Economies and Labour Markets

3.1 Introduction

> Who was to build the new buildings and motorways, make the castings, clean the cities, man the assembly lines, quarry the minerals, load the goods, bury the pipe lines? (Berger & Mohr, 1975: 122)

This sequence of rhetorical questions about an intrepid labour force capable of sustaining modern urban life appears in *A Seventh Man*, John Berger and Jean Mohr's remarkable portrait of Mediterranean migrant workers in the cities of northern and central Europe in the early 1970s. Combining photography, ethnography, statistics, poetry and Marxist political economy, Berger and Mohr trace the migrants' evolving relationship with the tentacular metropolis: from the seductive images of consumer goods relayed to their home villages to the apprehension as they transit through regional cities on the way to expatriation; from the bewilderment on arrival at their destination, through the physical and mental strains of their labour, to their eventual return home to flaunt the trappings of modern life and to entice the next in line to leave. Along this journey we encounter teams of Yugoslavian, Italian and Spanish workers deep beneath Geneva as they dig a new drainage system to allow the city to expand. The work is dangerous, exhausting and poorly paid. Barred from bringing their families with them, the workers lodge in wooden barracks on the other side of the city, where they spend most of their free time. These tunnel diggers, however, are not portrayed as helpless casualties of the international division of labour, but as rational decision-makers who 'undertake [the work] in order to earn and save the maximum amount of money in the shortest time' (Berger & Mohr, 1975: 163). Indeed, as Berger writes, 'time off [to go into town] is time prolonged and wasted' (ibid: 167).

As the book's title suggests, all the workers in *A Seventh Man* are male, something Berger openly acknowledges (ibid: 8). In August 1976, a year after the publication of Berger and Mohr's masterpiece, four East African Asian women and two men walked out of a photo-processing plant in Willesden, north-west London, in protest

N. Dines, *Migration and European Cities*, IMISCOE Research Series,
https://doi.org/10.1007/978-3-032-15379-1_3

at compulsory overtime and heavy-handed management. The next working day, they were joined by over 130 workers, mostly local women, two-thirds of whom were of South Asian origin. Demands quickly shifted to better wages and union recognition, triggering the start of a strike that would last two years. Although ultimately unsuccessful, the highly public dispute is seen as a landmark in British labour history, 'when the working class recognised the rights of women and minority workers to join a union as part of the British working-class movement' (McDowell, 2013: 151). The Grunwick strike, as it came to be known, overturned patriarchal stereotypes of South Asian women as homebound and docile. It also complicated images of ethnicity, migration and class in London. Many of the women had grown up in cities in East Africa, having left Kampala, Nairobi and Dar-es-Salaam in the 1960s and 1970s. Most came from middle-class backgrounds and had little previous experience of wage labour.

The managing director of the Grunwick plant was himself an immigrant. George Ward was born in Delhi in 1933 into a wealthy Anglo-Indian family that had become destitute by the time they moved to the British capital in 1948. Ward went on to train as an accountant in central London and spent three years working in Rio de Janeiro in the early 1960s before ploughing his savings into setting up Grunwick with two English friends he had met at his local church. Ward wrote his own version of the strike, entitled *Fort Grunwick*, in which he portrayed himself as a self-made man who, along with his loyal non-striking workers, was the victim of a politically motivated attack by the trade union movement (Ward, 1977). In doing so, he anticipated some of the central tropes of 1980s Thatcherite Britain.

A Seventh Man and the Grunwick strike present different facets of the economic relationship between migration and cities. Through their labour, the guest workers in *A Seventh Man* sustained the urban economies of northern Europe, yet they had little social or affective connection with the cities. In the case of the Grunwick strike, on the one hand we have migrant women who drew on family and community networks in an increasingly multicultural part of London to build support for the strike. On the other hand, we have the foreign-born entrepreneur who overcame personal hardship to build a business on the back of a cheap and—it was initially assumed—pliable workforce.

But the cities of the 1970s were also undergoing major changes. The slowdown in post-war growth in western Europe, together with the decline of Fordist modes of production and the crisis of the Keynesian model of state economic intervention, would give way to an era of capitalist restructuring characterized by an enhanced role for the private sector in macroeconomic policy. This shift had profound implications for the economic function, social composition, physical form and political governance of cities. Factory closures and the relocation of production, first to the urban periphery and later to lower-wage regions, left cities facing disinvestment, urban decay, rising unemployment and shrinking populations. Deindustrialization was often a drawn-out and uneven process, affecting some sectors and heavily industrialized cities more than others. In Central and Eastern European cities, the collapse of state socialism after 1989 was followed by 'shock therapy' structural adjustment programmes that included the closure of state industries, the removal of price controls, the abrupt end of secure forms of white- and blue-collar employment and a surge in emigration, although again these changes were not uniform and economic decline was often felt most acutely in smaller, mono-industrial centres (Smith et al., 2008).

In the wake of these economic and political conjunctures, new systems of production and consumption would emerge in Europe's cities as a result of, inter alia, the expansion of service and retail sectors, the development of new technologies, and the intensification of tourism and cultural industries. While some major metropolises would assume leading roles in an increasingly globalized economy, all cities have adopted more entrepreneurial forms of governance to varying degrees and with different levels of central government involvement, with the aim of attracting external investment (Harvey, 1989). Industrial cityscapes and class-differentiated neighbourhoods have been replaced or modified by the assorted trappings of regeneration and gentrification: from suburban retail complexes and flagship cultural institutions to renovated street markets and new nightlife districts. The emergence of a so-called 'post-industrial' urban society has been accompanied by the restructuring of government spending and the deregulation of labour markets. The decline of middle-income, male-dominated secure jobs in manufacturing has been offset by an increase in temporary and part-time forms of employment, but also by the permanent entry of increasing numbers of women into the workforce, previously engaged in precarious or unpaid work (McDowell, 2013).

Migration provides a lens through which to understand changes in urban capitalism in Europe. People who moved to cities since the 1980s have entered a very different context to that of the Spanish tunnel digger or the déclassé Asian Ugandan refugee of previous decades. Economic activities that were once a mainstay of employment and sociability have disappeared or lost their grip on urban life. Employment, welfare and pension rights have been eroded, while a free-market ideology that values self-reliance and flexibility over dependency and stability has gained ground across the continent. In many cities, especially in southern and eastern Europe, the vast majority of foreign nationals arrived after these changes had already begun and would contribute in different ways to the emergence of new urban economies.

One might therefore conclude that the two opening vignettes add little to a discussion of the economic links between migration and cities today. Yet the scenarios evoked by *A Seventh Man* and the Grunwick strike point to issues that remain central: the role of migrant labour in enabling cities to function, the gendered dimension of work, migrant entrepreneurship and industrial conflict. As such, they provide useful yardsticks for thinking about what has and has not changed, and counterbalance any assumption that our current urban condition marks a clean break with the past.

This chapter examines the ways in which international migration and urban economies interact, and the processes that shape this interaction.[1] Unlike the previous chapter, which focused primarily on the work of historians, this chapter draws on a

[1] The focus of this chapter is on international mobility. It is important to note, however, that internal labour migration to large cities remains a significant phenomenon and has increased in importance since the 2008 economic crisis and the exacerbation of regional disparities within countries, particularly in southern and eastern Europe. A telling example is the increased mobility of school teachers

range of disciplinary perspectives, including economic geography, labour sociology, as well as urban studies and migration studies. The chapter begins by looking at the role played by different migrant workforces in Europe's leading global cities, but also in smaller cities further down the urban hierarchy. It continues with a discussion of the gendered dimension of work and social reproduction and how such a focus reframes our understanding of the economic relationship between migration and cities, before turning to the role of migrant labour in the development of atypical forms of employment, particularly in informal and digital gig economies. The chapter concludes by examining the growth of migrant entrepreneurship in European cities over the last four decades and its contribution to the transformation of urban economies.

3.2 Migrant Labour Markets in European Global Cities

Since its emergence in the late 1980s, the idea of the 'global city' has provided a framework for understanding the economic links between migration and cities and, in particular, for highlighting the crucial role of migrant labour in the global economy. The term typically refers to cities that function as nodes of command and control in the global economy and have acquired a pivotal position vis-à-vis transnational flows of capital, people, goods and information. According to Saskia Sassen's 'global city hypothesis' (1991), the shift from manufacturing to services and the concentration of corporate headquarters in major cities exacerbated wealth inequalities and led to occupational and social polarization. The two 'ends' of the urban labour market—highly skilled professional and managerial positions on the one hand, and low-paid routine jobs servicing the urban elite on the other—attract and to a large extent depend on international newcomers. Indeed, a key contrast with the Swiss and German cities portrayed in *A Seventh Man* is that foreign-born migrants are no longer confined to the lower echelons of the workforce.

According to the various global city ranking systems, Europe's leading financial and business centre is London: the capital of one of the first industrialized nations, the former centre of a global empire, a long-established seat of international banking and a historic hub of international mobility (Eade, 2000). Paris, at the heart of another ex-colonial metropole, vies with London for the status of Europe's leading global city, while other major financial and political centres include Frankfurt, Brussels, Madrid, Milan and Amsterdam as well as Moscow and Istanbul. Despite their different post-war migration histories—mainly postcolonial in the case of Paris and London, guest worker in the case of Frankfurt and Brussels, and internal in the case of Madrid and Milan—these cities and their surrounding regions have experienced an increase

from southern Italy to cities in northern and central Italy, which not only reflects the high demand for an increasingly smaller pool of secure, permanent jobs, but also serves as a reminder that in many parts of Europe, public sector labour markets remain restricted to national citizens (Colucci & Gallo, 2017).

and diversification of international migration flows since the 1990s. Over the same period, the cities and their metropolitan hinterlands have also experienced some of the highest economic and demographic growth on the continent.

The movement of highly skilled migrants to major world cities is widely seen as a defining feature of globalization. Lars Meier (2014) identifies three reasons for the increase in professional migration in Europe: first, the marked growth in tertiary education in many countries since the 1960s; second, an increase in international mobility in the wake of the post-2008 economic recession among those who would otherwise remain in their home countries; and third, a widespread belief that intercity migration constitutes a necessary resource for career advancement. Professionals moving to and between Europe's top-tier cities cross borders relatively easily compared to other international migrants. Indeed, they are rarely publicly perceived as either migrants or workers, but rather, in the words of former Mayor of London Boris Johnson, as 'global talent' that city governments and businesses seek to attract (Raco & Kesten, 2018). While commonly seen as enhancing the diversity, competitiveness and attractiveness of cities through their economic contributions and cultural and consumer practices, the mobility of these migrant professionals also poses serious challenges to cities, for instance by increasing pressure on housing availability and accelerating processes of gentrification (White & Hurdley, 2003).

The popular image of a socially homogeneous, transient elite moving effortlessly between global cities is, however, somewhat simplistic. It overlooks differences in terms of gender, race and life stage, and ignores the fact that a growing number of professionals in Europe's major cities are from the global South (Back et al., 2018). Such factors can influence individuals' attachments to a particular place and their ability and willingness to move elsewhere (Yanasmayan, 2014). In his ethnography *Eurostars and Eurocities*, Adrian Favell (2008) reconstructs the stories of free-moving intra-EU professionals in London, Brussels and Amsterdam and considers local forms of exclusion, such as restricted access to social services and housing markets, 'that even the most global of cities can impose on these otherwise privileged foreign residents' and which ultimately hinder the possibility of a 'truly cosmopolitan, denationalized lifestyle' (Favell, 2008: xii). Favell also makes the important point that while Europe's middle classes have become more internationally mobile, they remain the exception rather than the rule. At the time of Favell's study, fewer than 1 in 50 western Europeans worked abroad, and most were concentrated in major international cities such as those at the centre of his research (ibid: 25). However, the situation changes when we move from the economic and political heartlands of the EU to southern and eastern Europe, where rates of international mobility are significantly higher, especially among young graduates who face precarious job prospects in their respective countries (King et al., 2016).

Early critical engagement with the global city hypothesis challenged Sassen's claims about occupational polarization in European cities (Body-Gendrot, 1996; Hamnett, 1994; Samers, 2002). It was argued that while this may have been the case in New York, it was not observable in London or Paris, where more generous welfare regimes and lower levels of labour immigration in the 1970s and 1980s

Occupation	1993/1994 (000s)	% foreign born labour	1999/2000 (000s)	% foreign born labour	2001/2002 (000s)	% foreign born labour	2004/2005	% foreign born labour
Chefs, cooks	29	**51**	34	**61**	31	**67**	38	**76**
Catering assistants	27	**42**	25	**52**	38	**55**	39	**62**
Care assistants	22	**n/a**	41	**48**	36	**38**	35	**56**
Cleaners	64	**41**	55	**46**	52	**61**	51	**69**
All occupations (London)	2,894	**25**	3,262	**27**	3,349	**30**	3,375	**34**
All occupations (UK)	24,449	7	26,687	**8**	27,114	9	27,599	**10**

Fig. 3.1 Total employment and the proportion of foreign-born labour, by occupation, in London, 1993–2005 *Source* Wills et al. (2010: 42)

limited the growth of low-wage service sectors. However, with the rise of international migration across Europe in the 1990s, the situation changed markedly. Pioneering research conducted in London during the 2000s revealed an emerging 'migrant division of labour' whereby a disproportionate number of the city's low-paid jobs in cleaning, hospitality and security were filled by foreign-born workers (May et al., 2007) (Fig. 3.1). Furthermore, the precarious or irregular legal status of many non-EU workers—for instance, those with student visas that restricted working hours—meant they were trapped in the least skilled and worst-paid sectors (Wills et al., 2010). The study stressed that these migrant workers were not just servicing the lifestyles of professional elites, but were playing a wider role in keeping London and its economy going, much like the migrant workers described by John Berger in *A Seventh Man.* This new migrant (and increasingly racialized) division of labour was not simply a consequence of London's global financial reach, but the result of decisions by the British state on labour market deregulation, welfare reform and immigration policy designed to secure Britain's competitive advantage in the global economy and to meet employers' demands for flexible, low-cost labour (May et al., 2007: 155–157).

This situation has not gone unchecked. From university service personnel in London to airport cleaners in Amsterdam, migrant workers in Europe's global cities have been involved in workplace disputes that have sometimes succeeded in securing better pay and conditions and the insourcing of services (Alberti & Però, 2018; Connolly et al., 2017). Like the Grunwick strike in the 1970s, these mobilizations have had a revitalizing effect on the wider labour movement. Not only have they spawned new grassroots unions and support alliances, but they have also managed to significantly shift public narratives about groups of workers who were previously far less visible and often considered unorganizable (Però, 2020).

In sum, global cities exacerbate regional disparities in wealth and are themselves highly unequal places. Foreign-born populations face far greater inequalities in income distribution than during the post-war era of mass labour migration. This said, migrants have moved into a far more complex spectrum of work than the theory of occupational segregation would suggest. Of the nearly one million newcomers who moved to London to work between 2001 and 2011, many, particularly the 'free movers' from the enlarged European Union, took up mid-level positions in health, education and administration, in part because of the greater transferability of qualifications.

3.3 Below the Top-Tier: Migrant Labour in Less Powerful Cities

Debates around global cities have been fundamental in redressing nation-centred accounts of migrant labour. Initially, most attention in Europe focused on command-and-control centres such as London, while smaller and peripheral cities were typically left out of the picture, even though they too were undergoing economic restructuring and were, to varying degrees, destinations for migration. According to Glick Schiller and Çağlar (2009), the focus of global cities research on selected cities and specific labour markets has hindered the formulation of a broader theory of locality in migration studies. Instead, the two anthropologists argue that all 'cities are now globalizing, but are embedded within differential power hierarchies and with varying outcomes' (ibid: 182). They call for a 'comparative scalar approach' to migrant settlement that addresses how migrants—through their labour, production of wealth and creation of social institutions and transnational connections—contribute to the repositioning of cities within national and global fields of economic and political power, be they 'top-scale' metropolises such as London and Paris or 'down-scale' deindustrialized and economically weaker cities with relatively limited prospects for employment and other economic activities.

A scalar approach is conducive to thinking about the types of labour opportunities in small and medium-sized cities and how these influence different pathways of migrant incorporation or, what Çağlar and Glick Schiller (2018: 20–21) have recently termed 'emplacement'.[2] Here I highlight three key areas: the restructuring of labour markets previously dependent on migrant workers; the expansion of migrant-dependent sectors typical of global cities into lower-tier cities; and the emergence of new sectors specific to smaller urban centres.

First, a scalar approach draws attention to the legacies of past migrations in relation to the restructuring of local labour markets. As discussed in Sect. 2.6, post-war labour migration was not exclusively associated with large cities, but also played a crucial role in mono-industrial cities such as the mill towns of northern England and the steel-producing centres of the Ruhr. Industries had recruited migrant workers in an attempt to maintain profitability in the face of growing international competition, but with the downsizing and factory closures that followed the price crises of the

---

[2] In their early writings, Çağlar and Glick Schiller used 'incorporation' as a way 'to theorize settlement outside the expectations of assimilation and the political values embedded in integration discourses.' Since 2018, they have opted for 'emplacement' because 'it emphasizes the ongoing processes of establishing social relationships within space and time and situates both migrants and non-migrants within these processes' (2018: 228).

1970s and changes in state industrial policy, many of these workers were abruptly laid off. Kalra (2000) examines the lives of Pakistani-Kashmiris in Oldham who, as young men, had worked night shifts to keep the local mills running and subsequently found low-paid employment or modest entrepreneurial opportunities in the taxi, retail and catering industries. The men's limited options were compounded by variables such as their older age and recent family reunification, which discouraged further migration, but also reflected the long-term decline of Oldham from a world-leading centre of textile production to a largely residential town characterized by high levels of deprivation on the edge of the Greater Manchester conurbation.

Second, a scalar approach encourages us to consider the extent to which labour markets conventionally identified with global cities exist elsewhere, and the conditions under which migrants access them. In the case of high-skilled migration, the enlargement of the European Union, the relaxation of restrictions on skilled workers from non-EU countries after 2000, and corporate relocation policies have all contributed to a more spatially dispersed professional mobility (Plöger & Becker, 2015). Local development strategies and marketing campaigns have attempted to take advantage of this expanding geography. Jaskułowski (2018) considers the case of Wrocław, the fourth most populous city in Poland. Since Wrocław became a secondary hub for multinational companies and a destination for workers from across Europe and Asia in the late 2000s, the local government has sought to stimulate development by promoting an image of a multicultural, tolerant city. Such a vision, however, does not match the reasons why professionals are attracted to Wrocław. Instead, living and working in the city is seen by many as a temporary career move, while for others it is an entry point into the European Union.

Similarly, the growing proportion of migrant workers in low-skilled sectors is not confined to major global cities, but has also occurred in smaller cities across Europe. The economic and political position of a city, the occupational circumstances of the local population, the consequences of labour mobility regimes and the strategic choices made by migrants themselves determine the extent and impact of a migrant workforce in different ways. In wealthy Nordic cities such as Oslo and Reykjavik, which are traditionally characterized by high levels of labour protection and low unemployment rates, the use of migrant labour has led to wage dumping in the hospitality and construction industries and labour market segmentation (Friberg et al., 2014). Migrant workers have also entered low-skilled labour markets in less affluent cities. Research on Newcastle-upon-Tyne in the mid-2000s highlighted an emerging division of migrant labour across a range of sectors, largely as a result of the free movement of Central and Eastern European nationals. Despite the fact that many migrants remained in the city only temporarily due to limited job prospects, regional authorities saw this labour supply as a stimulus to underperforming labour markets, even if it meant delaying investment in training programmes for local workers and under-utilizing the skills of migrants themselves (Stenning & Dawley, 2009).

Third, and finally, a focus on smaller urban localities allows us to shed light on migrant-dependent industries that are rarely, if ever, addressed in the global city literature, but which point to significant economic developments that arguably reorient our very understanding of the relationship between migration and cities. The industries in question are logistics, industrial food production and electronic contract manufacturing, and the cases considered here are located on the outskirts of three small cities, each with a population of around 100,000: Piacenza in northern Italy, El Ejido in southern Spain and Pardubice in the Czech Republic.

Piacenza's Le Mose logistics hub, used as a base by a number of multinational retailers including IKEA, was developed in the early 2000s and today employs around 2,000 people. Located 60 kms south-east of Milan at a junction of Italy's rail and motorway systems, it is a key node in the Po Valley logistics system, one of the most important distribution networks in southern Europe. Together with its counterparts in Bologna, Padova, Verona, Brescia and Milan, Le Mose functions as a sort of inland port and employs migrants, mainly from North Africa and South Asia, as porters through subcontracted cooperatives to ensure the smooth flow of goods (Cuppini et al., 2017; Cini & Goldmann, 2021).

El Ejido, located in the province of Almeria on the E15 motorway route linking Andalusia with Barcelona and Paris, is at the centre of Europe's largest greenhouse district that produces off-season fruit and vegetables for northern European supermarkets. Both the city and its labour-intensive agroindustry have grown rapidly since the 1990s thanks to an influx of casual, low-paid workers mainly from Morocco, Romania and sub-Saharan Africa, who now make up 30% of the city's population (Jiménez, 2010).

Finally, Pardubice, 100 kms east of Prague, has been home since 2000 to Europe's first Foxconn factory, the Taiwanese contract manufacturer of computers and other consumer electronics. Over the past twenty years, multinationals such as Foxconn have opened assembly plants in small, well-connected and low-wage cities in Central and Eastern European countries in order to gain direct access to EU markets. Since the Czech government introduced agency employment into the national labour code in 2004, companies have been able to take advantage of cheaper and non-unionized temporary migrant workers. Indeed, more than 60% of Foxconn's 5,000 workers are migrants from Romania, Bulgaria, Poland and Slovakia, recruited and transported from their countries of origin by international private work agencies and housed in hostels in Pardubice (Andrijasevic & Sacchetto, 2017).

The development of their respective industries has enabled all three cities to achieve strategic positions in global supply chains. This would not have been possible without access to a low-wage and flexible (if sometimes combative) migrant labour force. Migrants are active agents in the rescaling of Piacenza, El Ejido and Pardubice within global fields of economic power, but not in the way that Glick Schiller and Çağlar suggest. The conventional, marketable trappings of migration-driven urban diversity are largely absent. The homes of the workers at Le Mole are not concentrated in urban neighbourhoods, but scattered across the towns and villages of the Po plain. The migrant workers at Foxconn, like the tunnel diggers in *A Seventh Man*, live out their temporary stay between the factory floor and the hostel, having little interaction with the rest of Pardubice. By contrast, the settlement of day labourers in El Ejido and the surrounding shantytowns has long been a source of tension with local residents. Authorities in all three cities have not publicly promoted the presence of migrants as a source of economic growth. This is not simply a sign of political hostility or indifference to immigration, but reflects the limited influence of political actors in smaller localities in managing migration flows, regulating labour and defining local development strategies; competencies that have increasingly been taken over by multinational retail and manufacturing companies over the last twenty years.

3.4 Gendered Migration and Reproductive Labour

Migrant labour markets are not only differentiated by city type, skill level, nationality and race, but are also typically segmented along gender lines. The South Asian millworkers in Oldham were all men, as are today's porters in the logistics hub of Piacenza. In sectors where both men and women are employed, jobs may still be highly gendered. For instance, research on migrant cleaners in London found that women were numerically dominant in semi-private spaces such as hotels, while men dominated in semi-public spaces such as offices and public transport (May et al., 2007: 159). As a result of the sharp increase in female labour migration since the 1980s (Kofman & Raghuram, 2015), women now make up the majority of migrant workers in many European cities (ILO, 2021). Gender intersects with other social divisions such as race and class, which is bluntly reflected in the disproportionate number of migrant women, particularly migrant women of colour, in low-paid and unpaid work (McDowell, 2013). At the same time, gender reconfigures the very way we interpret the economic, social and political relationships between migration and cities.

The compelling argument that migrant workers keep urban economies running has already been made in relation to low-paid service jobs such as office cleaning (May et al., 2007). But migrants, and particularly migrant women, are increasingly crucial to the reproduction and maintenance of society and social relations more generally (Rigo, 2022). Social reproduction can be understood as the array of mental, manual and biological activities, performed predominantly by women in private and public settings, that sustain both life and labour power on a daily basis and reproduce the next generations (Laslett & Brenner, 1989). Activities include adult and child care, food preparation, child socialization, maintaining community ties, emotional support and sex work. As feminist urban scholars Peake et al. argue, ‘it is not possible to think through the urban without considering the role and relations of social reproduction [and] the interdependence between relations of social reproduction and production’ (Peake et al., 2021: 1–2).

Attending to social reproduction therefore means broadening our understanding of urban inequalities to include the different conditions under which social life is reproduced in cities. It extends John Berger's opening question about who buries the pipelines to ask whose labour makes the labour of others possible. In short, it means reimagining how we conceptualize the economic organization of cities. Most reproductive work depends on sustained bodily contact: while it can be outsourced (for example, from the family to the market), it cannot be delocalized and thus stands in contrast to the narratives of fluidity and immateriality that have been used to depict the transition to globalized post-industrial societies (McDowell & Dyson, 2011). With the increasing marketization of care provision and the entry of greater numbers of native women into the workforce, migrant labour has increasingly performed the activities that sustain urbanization. In addition to ensuring the social reproduction of their employers/clients, migrant workers also organize the reproduction of their own households in countries of destination or origin. The 'care chains' that result from this 'international division of reproductive labour' (Parreñas, 2000, 2012) contribute to the formation of translocal ties and, as such, play a fundamental role in globalization from below.

Domestic work is a key segment of the reproductive labour market that has expanded rapidly around the world since the 1980s. In Europe, this sector is largely, though not exclusively, occupied by migrant women and is characterized by some of the lowest paid, most precarious, physically and emotionally demanding work in cities (Anderson, 2000; Lutz, 2008, 2011). The reasons for the increased demand for migrant labour are diverse and vary from country to country. They include the aforementioned increase in the full-time labour participation of local women (the majority of whom were previously responsible for reproductive tasks at home), the ageing of the population and changes in intergenerational relations, the restructuring of public services, and immigration policies aimed at regularizing care workers (Marchetti, 2022: 22–28). The models of welfare regimes (Esping-Andersen, 1990), from family-based care systems in southern Europe to liberal and social democratic state-centred types of assistance in northern Europe, and the various transitions they have undergone in recent decades, have strongly influenced the emergence of different domestic labour markets, from the prevalence of live-in workers directly employed by families in Italy to a 'migrant in the market' model of employment based on care agency staff in the UK (van Hooren, 2012).

The extensive literature on migrant domestic work in Europe includes studies in single cities (Anderson, 2001; Näre, 2011; Pérez & Stallaert, 2016), multi-sited research in single countries (Lutz, 2011), as well as global comparative analyses that include European cities (Lindio-McGovern, 2012). In some of these works, the city represents little more than a research site, and there is limited interest in considering the significance of the urban scale or the relationship between domestic work and urbanization processes. Other studies, especially those using ethnographic approaches, have explored the spatial dimensions of domestic labour, shedding light on the nuanced power relations and gender roles that shape migrants' working lives in cities. Ester Gallo and Francesca Scrinzi's research on male Asian porters in upper-middle-class condominiums in Rome discusses how this niche of reproductive labour combines feminized nurturant tasks, such as cleaning and caring, which take place primarily in the private spaces of the home, and less relational, masculinized tasks, such as maintenance and security, which are performed in the semi-public spaces of the buildings and on the surrounding streets (Gallo & Scrinzi, 2019). Various scholars have considered how urban public spaces provide domestic workers with support networks, information hubs and informal systems of labour exchange (Cvajner, 2019; Dines, 2012). In such cases, urban space becomes a site where reproductive labour is itself reproduced.

Focusing on the interconnections between reproductive labour, gendered migration and urbanization represents an exciting frontier for future research and has the potential to overturn commonplaces about migration and cities. In her study of post-Soviet domestic workers in Trento in the Italian Alps, Martina Cvajner (2019) describes how in the 1990s scores of east European women first migrated to Naples due to the abundance of live-in domestic work and a range of self-organized transport and postal services. Many then moved north as soon as they could, often to smaller provincial towns, in search of better conditions (Cvajner, 2019). Thus, if we look at urban-bound migration through the lens of reproductive labour, it is Naples—a city that has been marginal to accounts of epochal urban change—rather than London, Paris or Amsterdam that functions as a 'gateway city.'

3.5 Non-standard Employment in Informal and Digital Gig Economies

In addition to the low-wage care labour markets, international migrants—both male and female—are overrepresented in atypical forms of employment that have proliferated under neoliberal capitalism. Here I consider two areas: informal economic activities, which were the subject of considerable academic discussion in the 1990s and early 2000s, and digital gig work, which has been the focus of a burgeoning area of research since the late 2010s. Informal and non-standard forms of employment can provide newcomers, especially those with irregular status, with access to labour markets and potential stepping stones to better jobs, but they often perpetuate insecurity, increase inequality and block upward mobility (ILO, 2018). The intertwining of migrant labour and atypical work is most evident at the urban scale, and the intensity and significance of this relationship provides important insights into the position of a particular locality—be it a neighbourhood, city or metropolitan region—within national and global hierarchies of economic and political power.

The informal economy essentially refers to economic activities that exist outside the regulatory control of the state. These activities can be subdivided into two general types: on the one hand, unregistered private enterprises, and, on the other, undeclared work without legal and social protections that can be found in either the formal or the informal sector (Williams, 2023). The adjective 'informal' and related terms such as 'shadow', 'grey', 'hidden' and 'underground' are often used to portray such activities as negative, deficient or residual, in contrast to the norm represented by the 'formal' economy. However, scholars have long emphasized the ambivalent nature of economic informality, which, depending on the context, can be as much about creativity, self-organization and diversification of household income as it is about exploitation, abuse or evasion (Hart, 1973; Portes et al., 1989).

After the 1980s, academic interest in the informal economy in the West was based on the view that the sector was expanding (Samers, 2004). Among the various arguments, two explanations stood out. First, on the demand side, the growth of informal economic activity was interpreted as a consequence of economic restructuring and labour deregulation, which was seen to be particularly acute in global cities. The need to compete with established service providers, to perform outsourced activities and to meet the consumption needs of low-income workers led to the rise of 'downgraded' firms, sometimes established by migrants or people with a migration background and operating partly or entirely outside institutional frameworks (Sassen, 1996). Second, on the supply side, the growth of international migration to large cities was seen as an important (if not the only) source of cheap labour that favoured the development of the informal sector (Portes et al., 1989). In their analysis of Brussels between the 1960s and the late 1990s, Christian Kesteloot and Henk Meert (1999) found that informal economic activity varied according to the characteristics of an area and its migrant population. In one central neighbourhood, informal business and employment practices were associated with a flourishing catering industry that had drawn on the dense social networks of a specific migrant group and the rising demand of the general population, while in another neighbourhood, informal activities consisted mainly of survival strategies that reflected higher levels of poverty and the absence of a numerically dominant group. Informality not only provided pathways of incorporation for migrants in the Belgian capital, but was also an important driver of urban growth. The long opening hours of fast-food outlets, together with the off-the-books employment of undocumented co-nationals, the unpaid labour of family members and the non-compliance with mandatory closing days, all contributed to the development of the city's night-time economy.

At the turn of the millennium, several studies converged on the argument that the availability of foreign labour was not the underlying cause of the growth of informal employment, but rather that an increasing proportion of new migrants found precarious work in sectors such as care, retail and construction that had already been heavily deregulated in previous years (Reyneri, 1998; Samers, 2004; Wilpert, 1998). In addition, the informal economy had a long history in some European cities. In eastern Europe, the 'second' economies and informal marketplaces that had developed during the Soviet era continued after the 1990s, providing entrepreneurial and employment opportunities for migrants moving to cities in the region (Sik & Wallace, 1999; Turaeva, 2013). In major southern European cities such as Athens, Naples and Lisbon, informal work practices accounted for up to 25% of the local labour market, reflecting the historical importance of sectors with low levels of protection, such as small-scale manufacturing and construction (Baganha, 1998; Leontidou, 1990; Reyneri, 1998). The prospect of immediate income through informal channels and the perception that these enjoyed broad social acceptance and a degree of state acquiescence, were motivating factors for migration to these new urban destinations in the 1980s and 1990s (Dines, 2012; Schmoll & Semi, 2013; Vaoiu, 2002).

Academic interest in informal economic activities in European cities has been sporadic over the last two decades. Many aspects previously associated with migrants' informal employment have since been subsumed under the umbrella label of precarity (Han, 2018), while critical engagement with informality itself has largely focused on the global South (Roy, 2005). The subsequent shift in perspective, whereby informality is no longer seen in terms of deficiency and marginalization, but as an organizing logic that produces value, has informed some recent reflections on European contexts. An in-depth qualitative study of Pakistani migrant and settled workers in Sheffield urges researchers to consider 'a range of predominantly social and cultural motives [that] help explain the continued normality of informal work within urban settings' (Rodgers et al., 2019: 476). The authors suggest that informal work is about more than economics and constraints: it provides migrant and minority workers with 'opportunities, even status, giving them agency in what would otherwise be a disempowered situation in an urban economy' (ibid: 491).

Of course, employment in the 'formal' economy does not guarantee protection from precarity. In recent decades, economic deregulation has led to increasingly flexible labour markets, while advances in digital technology have facilitated the expansion of on-demand work arrangements. In particular, app-based services such as ridesharing, food delivery and domestic cleaning have led to the spatial and social reorganization of urban labour markets. Migrant workers have played a decisive role in these processes (Altenried, 2024; Gebrial, 2024; van Doorn & Vijay, 2024; van Doorn et al., 2023). Platform companies in major European cities have benefited from a ready supply of migrant labour to rapidly scale up their businesses and compete with rivals. Through algorithmic management, platforms control their workforce at arm's length and are able to respond to fluctuating demand at almost no cost, as workers are only paid when they work (Altenried, 2024). The fact that many are formally self-employed means that the responsibility for the social reproduction of labour falls entirely on the shoulders of workers.

For migrants, digital platforms constitute 'plug-and-play arrival infrastructures' (van Doorn & Vijay, 2024: 1143). In other words, they offer relatively easy entry points into the labour market: little training or language skills are required, and even undocumented migrants can overcome legal and bureaucratic hurdles by sharing or renting other workers' official accounts. The downside is low-paid, exhausting and often hazardous employment that traps migrants in working poverty. Many are forced to work simultaneously on different platforms or in other sectors, leading not only to a drastic increase in work hours, but also to a blurring of work and leisure. The precarity associated with platform work is compounded by the unpredictable life cycles and deterritorialized loyalties of gig companies. Moritz Altenried's embedded ethnography of food delivery couriers in Berlin, for example, considers the consequences of Deliveroo's unexpected exit from the German market. For the city's 1,000 riders, mainly from the Middle East and austerity-hit southern European and Latin American countries, this meant the sudden loss of their main or only source of income and the prospect of underemployment, competition for places on other platforms, or return migration (Altenried, 2024). As in other low-wage sectors, there have been a number of translocal campaigns to improve conditions, although collective representation has been hampered by the geographical dispersion of workers and their formal status as independent contractors (Borghi et al., 2021).

Digital gig economies have contributed significantly to the reorganization of business and employment practices in large cities, and to the transformation of urban life more generally. They have been described as the first 'planetary labour market' (Graham & Anwar, 2020), with researchers reporting cases of migrants working for

	Country of origin (%)		Gender (%)		Age (%)		
	Native	Non-native	Male	Female	17-29	30-49	>50
Copenhagen	10.0	90.0	20.0	80.0	58.0	38.0	4.0
Paris	8.0	92.0	12.0	88.0	20.0	70.0	10.0
Berlin	0.0	100	42.0	58.0	54.0	46.0	0.0
Rome	56.0	44.0	4.0	96.0	18.0	44.0	38.0
Stockholm	16.0	84.0	30.0	70.0	58.0	36.0	6.0

Fig. 3.2 Breakdown of the origin, gender and age of workers for the online care service provider Yoopies, based on a dataset of 250 profiles (50 from each city) *Source* Giuliani and Parciani (2025). *Note* The terms 'country-of-origin', 'native' and 'non-native' are cited from the original source. In their text, Giuliani and Parciani use 'non-native' interchangeably with 'ethnic', 'race' and 'migrant' and so it is not clear if this category includes locally-born residents with migration backgrounds

the same companies in cities in different countries and even on different continents (Altenried, 2024). At the same time, online platforms reconfigure existing forms of contingent labour and reproduce stratified markets traditionally characterized by racialized and gendered workforces (van Doorn & Vijay, 2024). It is important to note, however, that the digital gig economy is far less dependent on foreign migrant labour in smaller or poorer European cities, or in many parts of southern and eastern Europe. For example, despite the sizeable migrant population in Naples,[3] the vast majority of the city's 2,000 platform riders are local people with few alternatives to earn an income (Iuliano, 2022), in contrast to couriers in central and northern Italian cities, who are largely of migrant origin. Meanwhile, data on workers employed via an online care service provider platform in five European capital cities (Fig. 3.2) indicates a 100% 'non-native' workforce in Berlin, but a relatively high 'native' presence in Rome. This is largely because migrant care workers in the Italian capital rely far more on informal offline networks to find employment (Giuliani & Paraciani, 2025). Thus, digital gig economies not only reconfigure labour relations, but also shed further light on the different, context-dependent pathways of migrant emplacement in European cities. These pathways reflect not only the consequences of migration policies on economic livelihoods but also the particular position of localities within multiscalar networks of economic and political power (Çağlar & Glick Schiller, 2018).

3.6 Migrant Entrepreneurship and Urban Development

In his essay 'Immigration and the Metropolis', the Cuban-American sociologist Alejandro Portes challenges the global city literature's emphasis on occupational segregation by observing: 'the attraction that the large metropolis exercises on migrants is multidimensional. When employment opportunities do not readily present

[3] According to 2022 statistics, Naples had 53,000 resident foreign nationals, making it the sixth Italian municipality in terms of migrant population after Rome (339,000), Milan (254,000), Turin (125,000), Bologna (59,000) and Genoa (58,000) (https://www.tuttitalia.it/campania/59-napoli/statistiche/cittadini-stranieri-2022/).

themselves, immigrants create them' (Portes, 2000: 157). While the decision to enter self-employment often arises as a result of exclusion from labour markets or an unexpected change in circumstances, it can also be the intended outcome of a migratory project. The closure of post-war labour recruitment programmes in the 1970s and the subsequent restructuring and internationalization of urban economies have led to a significant increase in migrant entrepreneurship across Europe. In Germany, for example, the number of migrant enterprises almost tripled from the 1990s to reach 750,000 in 2014 (Räuchle & Schmiz, 2019), while in major cities such as Vienna, Berlin, Zurich, London and Amsterdam, migrants account for around a third of all self-employed people (Parzer & Huber, 2015).

Over the past three decades, migrant entrepreneurship has become an established focus of research on migration and cities (Aytar & Rath, 2012; Çağlar & Glick Schiller, 2018; Jones & Ram, 2024; Rath, 2000). An early goal was to understand the relevance of ethnicity to the formation, survival and success of migrant businesses, such as the role of community networks in providing financial support, or the concentration of fellow group members in a particular area that made the provision of certain products or services feasible. Pioneering research in the United States coined the term 'ethnic enclave' to refer to urban areas with clusters of migrant populations and businesses that primarily served customers from the same cultural or national background (Wilson & Portes, 1980). However, many migrant-owned businesses have no discernible link to ethnicity. Some may cater to the needs of all migrants, as in the case of money transfer services, others may not target them at all, while still others may be indistinguishable from 'mainstream' businesses and located outside areas with high migrant populations. An emphasis on 'ethnic entrepreneurship' may also risk reinforcing stereotypes that certain groups are more likely to be self-employed than others. Indeed, a body of literature has explored how migrants strategically negotiate, perform and reinvent ideas about ethnic difference for business purposes (Deng, 2020; Hall, 2012). The term 'migrant entrepreneurship' is therefore preferred as a broad umbrella term for the range of business activities established by foreign-born migrants in cities, from micro-enterprises, such as market stalls, fast-food restaurants and general stores, which are relatively easy to set up but whose owners are often rooted at the lower end of the market, to larger-scale businesses such as garment factories, which potentially make migrant groups key players in urban and regional economies.

In response to the supply-side and ethnicity-focused explanations that characterized early studies of migrant entrepreneurship in North America, scholars in Europe from the 1990s onwards turned their attention to the complex interplay between migrant agency, structural dynamics and institutional frameworks. In particular, Dutch researchers working on Amsterdam developed the 'mixed embeddedness' approach, which combined the micro-level of entrepreneurs' resources, networks and actions with the meso-level of urban economic conditions and the macro-level of legal, policy and regulatory frameworks to understand how different opportunity structures shaped self-employment trajectories and facilitated or hindered the growth of particular markets (Kloosterman & Rath, 2001; Kloosterman et al., 1999).

This interactive approach encouraged researchers to question the common assumption that migrant and 'native' entrepreneurship were separate phenomena (while still considering specific issues such as racism and discrimination in labour markets). It also heightened attention to changes in opportunity structures both over time and at different levels and scales, and enabled systematic comparative research across different localities (Kloosterman & Rath, 2018).

Opportunity structures for migrant entrepreneurship vary not only between countries, but also between and within cities. They are shaped by, among other things, different consumer practices, market trends, zoning laws, development agendas, business regulations and translocal trade networks, all of which can be influenced by the economic activities of migrants themselves. The economic, cultural and political power of some leading global cities such as London, Paris and Amsterdam, and the density of migrant networks and organizations within them have enabled the development of new service industries for a diverse clientele, alongside the maintenance of residual retail sectors abandoned by local or corporate competitors (Glick Schiller & Çağlar, 2009; Sassen, 1996). Other, more economically peripheral cities, including those not historically associated with high levels of international migration, have sometimes become important centres of migrant entrepreneurship due to their strategic location, accessibility and low levels of local competition. This was the case in the 1990s for a number of Central and Eastern European cities, such as Budapest and Belgrade, which acted as gateways for Chinese shuttle traders thanks to short-lived periods of visa-free entry and low import tariffs to encourage foreign investment, the post-socialist demand for cheap commodities, and the globalization of Chinese migration and trade (Milutinovic, 2008; Nyíri, 2002). These same cities subsequently became regional distribution centres for Chinese goods. A striking example is the Block 70 shopping mall in the Novi Beograd municipality of Belgrade, which has become the base for around 1,200 Chinese traders and attracts buyers from surrounding countries (Korac, 2013).

Recent research has looked at variation within the same cities. For example, a study of different clusters of migrant businesses in crisis-stricken Athens identified a changing set of factors such as the number of closed shops on a street, the cost of rent, the pre-existing reputation of a neighbourhood, the residential trajectories of migrants, and proximity to public transport (Polyzou, 2020). Scholars have also turned their attention to smaller cities, where, for example, limited municipal support and buoyant labour markets that discourage self-employment are offset by close-knit bonding networks that support spatially dispersed migrant businesses serving a non-migrant clientele (Räuchle & Schmiz, 2019). Taken together, this scholarship reaffirms the importance of developing a context-sensitive and multidimensional analysis of migrant entrepreneurship and warns against the pitfalls of generic city-type models that make a priori assumptions about the importance of city size or intra-urban location (ibid).

As well as looking at the opportunity structures of different cities, researchers have been keen to understand how migrant entrepreneurship can potentially contribute to urban and economic growth. In addition to direct job creation, which can reach significant levels in some national and urban contexts,[4] some migrant businesses, such as 'ethnic' restaurants, are seen to enhance diversity, thereby attracting new investment, customers and visitors to urban localities, while others, such as wholesale businesses, are able to reposition cities within intercontinental supply chains (Jones et al., 2019; Marouda, 2020). As such, migrant entrepreneurs contribute to the diversification and rescaling of urban economies, which can result in public recognition for certain migrant groups and ultimately reorient local development policies (Glick Schiller & Çağlar, 2009).

The effects of migrant entrepreneurship are not linear or incremental, nor are they necessarily beneficial to the wider migrant and non-migrant population. In their work on Halle, a 'disempowered' medium-sized city in eastern Germany, Çağlar and Glick Schiller (2018) consider how migrant-run shops played a key role in revitalizing the city centre after German reunification by providing affordable goods and services to financially struggling residents, but were among the first to be replaced by chain stores as urban regeneration progressed. Scholars have also explored the ways in which businesses, particularly in the catering sector, are implicated in processes of 'ethnic branding', place marketing and gentrification, which in turn can lead to the displacement of migrant entrepreneurs and residents alike as rents rise and corporate investment flows into once-neglected neighbourhoods (Kadıoğlu Polat, 2020; Parzer & Huber, 2015; Shaw et al., 2004; Stock & Schmiz, 2019). The commodification of ethnicity is most acute in major metropolises with their various 'Chinatowns,' 'Little Istanbuls' and 'Curry Miles', but has increasingly become a feature of medium-sized cities across Europe, where migrant-run eateries are widely perceived to enhance the attractiveness of otherwise homogeneous urban environments (see, for example, Čapo & Kelemen, 2018 on Zagreb). In reality, many small migrant businesses operate at a subsistence level and are extremely vulnerable to external pressures, although there is evidence that proactive and creative strategies to mitigate harsh conditions—such as 'patch-working', reliance on supplementary forms of income—also enable business growth (Villares-Varela et al., 2018). As in-depth research in the West Midlands has illustrated, 'survival-oriented' enterprises still manage to create jobs and provide significant social benefits, cushioning the incorporation of new arrivals and sustaining retail and service provision in their local areas (Jones et al., 2019).

Much of the academic literature on migrant entrepreneurship has tended to focus on small retail businesses in less affluent neighbourhoods. However, with the changing nature of urban economies, some international migrants have become involved in larger scale and higher value-added activities (Jones & Ram, 2024; Kloosterman & Rath, 2018). An illustrative case is the revitalization of the textile and garment industry in the medium-sized Tuscan city of Prato by Chinese migrants (Ceccagno, 2015, 2017; Lan, 2015). In the late 1980s, Chinese entrepreneurs began to set up supplier firms in dilapidated factory spaces rented from locals, initially

[4] In Germany, for example, migrant-owned businesses employed 2.2 million people in 2014, or 18 per cent of the total workforce in the private business sector (Räuchle & Schmiz, 2019: 1768).

relying on modest capital and basic sewing skills, employing co-nationals and operating on very low profit margins. In the years that followed, Chinese production transformed Prato's ailing industrial district, which had struggled to compete in the global market, into a centre for fast fashion production. As a result, Prato now has one of the highest concentrations of Chinese nationals in Europe, accounting for 17% of the city's 200,000 inhabitants (https://demo.istat.it/).

Sociologist Antonella Ceccagno identifies two key factors behind this rapid growth. First, between 1990 and 1998, Italy imposed a ban on new entrepreneurship for immigrants from countries, including China, where Italians were prohibited from self-employment, resulting in a concentration of Chinese workers in Prato and existing businesses filled with 'frustrated potential micro-entrepreneurs' (Ceccagno, 2015: 1114). The lifting of restrictions led to the rapid opening of small firms and expansion into new areas of production. Second, growth was based on a production regime in which employers and employees worked and slept together in the factories during periods of peak demand. This allowed for just-in-time production and what Ceccagno calls 'in situ offshoring' to meet the short development cycles, rapid prototyping and small batch sizes required by the apparel industry.

The unintended consequences of restrictive immigration policy and the radical reconfiguration of spaces of production and reproduction meant that Chinese entrepreneurs and workers were able to create 'almost from scratch the largest migrant-run fast fashion centre in Europe' and contribute to 'reducing the downscaling of [Prato]' (ibid: 1113, 1125). This process has not been without controversy, from public scandals about poor labour conditions in Chinese factories to accusations of unfair competition and the rise of anti-Chinese sentiment among sections of the Italian population. Chinese migrants have also failed to reverse the downward spiral of the native-run textile industry. Nevertheless, and despite the various crises in the fast fashion sector itself, Prato is now at the centre of an integrated global value chain: traders source material from China and Turkey, while buyers flock to Prato from Europe and beyond. In other words, the combination of international migration and capital has fundamentally transformed Prato's economy, changing both the nature of the industrial sector and the structure of the labour market.

3.7 Conclusion

This chapter has traced the changes in the connections between international migration and urban economies through the body of research that has developed since the late 1990s. Clearly, the general situation has changed since the time of Berger and Mohr's tunnel workers in Geneva and the Grunwick strike in north London. The urban economies of all cities, including smaller, peripheral and disempowered cities, have been restructured by the increasing globalization of economic relations, the emergence of new sectors, the multiplication of types of migration, the diversification of labour markets and the growth of migrant entrepreneurship. The current overrepresentation of international migrants among the self-employed in many European cities, for example, reflects shifts from labour-intensive industries to service-based economies (Jones & Ram, 2024). Indeed, it is no coincidence that many of the early Asian-owned small businesses in the UK were set up by people laid off from industrial jobs in urban centres such as Oldham which had previously been destinations for labour migration.

Opportunities for employment and self-employment are a key motivation for migrants to move to and through cities. As shown, the incorporation of newcomers into local economies is never a smooth process, but takes place in a context of racism, poor working conditions and legal restrictions on labour market participation, among others. The city, however, is never simply the 'natural' setting for migrants' economic activities. As the important work of Glick Schiller and Çağlar demonstrates, migrants actively contribute to the reconfiguration of urban capitalism and the rescaling of cities within broader networks of economic power. It is precisely through the economic dimension that these two scholars have advanced theoretical thinking on the migration-city nexus. At the same time as recognizing the constitutive role of migrant labour and entrepreneurship in urban restructuring and how these are structured by international migration regimes, it is also necessary to *de-migrantize* our analysis in order to take into account the diversity of economic trajectories within migrant groups, the stark differences between European cities, and the broader shifts in the relations of production and social reproduction among urban populations as a whole.

Nevertheless, the vignettes that opened this chapter point to some enduring themes: racialized and gendered divisions of labour; the workplace as a site of conflict; the instrumental role of labour in long-term migration projects. Perhaps most significant of all, despite the increasingly interconnected global economy, migration remains crucial to the functioning of cities, from burying the pipelines in the 1970s to reproducing the workforce and social life today.

References

Alberti, G., & Pero, D. (2018). Migrating industrial relations: Migrant workers' initiatives within and outside trade unions. *British Journal of Industrial Relations, 5*(4), 693–715.

Altenried, M. (2024). Mobile workers, contingent labour: Migration, the gig economy and the multiplication of labour. *Environment and Planning A: Economy and Space, 56*(4), 1113–1128.

Anderson, B. (2000). *Doing the dirty work? The global politics of domestic labour*. Zed Books.

Anderson, B. (2001). Different roots in common ground: Transnationalism and migrant domestic workers in London. *Journal of Ethnic and Migration Studies, 27*(4), 673–683.

Andrijasevic, R., & Sacchetto, D. (2017). "Disappearing workers": Foxconn in Europe and the changing role of temporary work agencies. *Work, Employment and Society, 31*(1), 54–70.

Aytar, V., & Rath, J. (2012). *Selling ethnic neighbourhoods. The rise of neighbourhoods as places of leisure and consumption*. Routledge.

Back, L., & Sinha, S. with Bryan, C., Baraku, V., & Yembi, M. (2018). *Migrant city*. Routledge.

Baganha, M. I. (1998). Immigrant involvement in the informal economy: The Portuguese case. *Journal of Ethnic and Migration Studies, 24*(2), 367–385.

Berger, J., & Mohr, J. (1975). *A Seventh Man. A book of images and words about the experience of migrant workers in Europe*. Penguin.

Body-Gendrot, S. (1996). Paris: A "soft" global city? *Journal of Ethnic and Migration Studies, 22*(4), 595–605.

Borghi, P., Murgia, A., Mondon-Navazo, M., & Mezihorak, P. (2021). Mind the gap between discourses and practices: Platform workers' representation in France and Italy. *European Journal of Industrial Relations, 27*(4), 425–443.

Çağlar, A., & Glick Schiller, N. (2018). *Migrants and city-making: Dispossession, displacement and urban regeneration*. Duke University Press.

Čapo, J., & Kelemen, P. (2018). Mixing ethnic and non-ethnic economic strategies: Migrant entrepreneurs in Zagreb. *Narodna Umjetnost: Hrvatski Časopis Za Etnologiju i Folkloristiku, 55*(2), 29–56.

Ceccagno, A. (2015). The mobile emplacement: Chinese migrants in Italian industrial districts. *Journal of Ethnic and Migration Studies, 41*(7), 1111–1130.

Ceccagno, A. (2017). *City making and global labor Regimes: Chinese immigrants and Italy's fast fashion industry*. Palgrave Macmillan.

Cini, L., & Goldmann, B. (2021). The worker capabilities approach: Insights from worker mobilizations in Italian logistics and food delivery. *Work, Employment and Society, 35*(5), 948–967.

Colucci, M., & Gallo, S. (Eds.). (2017). *In cattedra con la valigia: Gli insegnanti tra stabilizzazione e mobilità. Rapporto 2017 sulle migrazioni interne in Italia* (In the classroom with a suitcase: Teachers between stabilization and mobility. 2017 Report on internal migration in Italy). Donzelli.

Connolly, H., Marino, S., & Lucio, M. M. (2017). "Justice for Janitors" goes Dutch: The limits and possibilities of unions' adoption of organizing in a context of regulated social partnership. *Work, Employment and Society, 31*(2), 319–335.

Cuppini, N., Frapporti, M., & Pirone, M. (2017). Logistics struggles in the Po Valley region: Territorial transformations and processes of antagonistic subjectivation. *South Atlantic Quarterly, 114*(1), 119–134.

Cvajner, M. (2019). *Soviet signoras: Personal and collective transformations in eastern European migration*. University of Chicago Press.

Deng, T. (2020). Chinese immigrant entrepreneurship in Italy's coffee bars: Demographic transformation and historical contingency. *International Migration, 58*(3), 87–100.

Dines, N. (2012). *Tuff city: Urban change and contested space in central Naples*. Berghahn.

Eade, J. (2000). *Placing London: From imperial capital to global city*. Berghahn.

Esping-Andersen, G. (1990). *The three worlds of welfare capitalism*. Polity Press.

Favell, A. (2008). *Eurostars and Eurocities*. Blackwell.

Friberg, J. H., Arnholtz, J., Eldring, L., & Hansen, N. W. (2014). Nordic labour market institutions and new migrant workers: Polish migrants in Oslo, Copenhagen and Reykjavik. *European Journal of Industrial Relations, 20*(1), 37–53.

Gallo, E., & Scrinzi, F. (2019). Migrant masculinities in-between private and public spaces of reproductive labour: Asian porters in Rome. *Gender, Place & Culture, 26*(11), 1632–1653.

Gebrial, D. (2024). Racial platform capitalism: Empire, migration and the making of Uber in London. *Environment and Planning A: Economy and Space, 56*(4), 1170–1194.

Giuliani, G. A., & Paraciani, R. (2025). Contextualizing inequalities in the gig economy: Evidence from online cleaning platforms in five European cities. *International Journal of Sociology and Social Policy*. https://doi.org/10.1108/IJSSP-12-2024-0619

Glick Schiller, N., & Çağlar, A. (2009). Towards a comparative theory of locality in migration studies: Migrant incorporation and city scale. *Journal of Ethnic and Migration Studies, 35*(2), 177–202.

Graham, M., & Anwar, M. A. (2020). The global gig economy: Toward a planetary labor market. In. A. Larsson & R. Teigland (Eds.), *The digital transformation of labor: Automation, the gig economy and welfare* (pp. 213–234). Routledge.

Hall, S. (2012). *City, street and citizen: The measure of the ordinary*. Routledge.
Hamnett, C. (1994). Socio-economic change in London: Professionalization not polarization. *Built Environment, 20*(3), 192–203.
Han, C. (2018). Precarity, precariousness, and vulnerability. *Annual Review of Anthropology, 47*(1), 331–343.
Harvey, D. (1989). From managerialism to entrepreneurialism: The transformation in urban governance in late capitalism. *Geografiska Annaler. Series B, Human Geography, 71*(1), 3–17.
Hart, K. (1973). Informal income opportunities and urban employment in Ghana. *The Journal of Modern African Studies, 11*(1), 61–89.
ILO (International Labour Organization). (2018). Informality and non-standard forms of employment. https://www.ilo.org/wcmsp5/groups/public/---dgreports/---inst/documents/publication/wcms_646040.pdf
ILO (International Labour Organization). (2021). *ILO global estimates on international migrant workers: Results and methodology*. ILO.
Iuliano, V. (2022, June 19). Riders "fantasma": A Napoli più di 2mila ma solo uno su dieci ha contratto e tutele ("Ghost" riders: More than two thousand in Naples but only one in ten has a contract and protection). *Il Mattino*. https://www.ilmattino.it/pay/edicola/napoli_riders_contratto_tutele-6761877.html
Jaskułowski, K. (2018). A study of local incorporation of migrant professionals in Wrocław. *International Migration, 56*(5), 123–136.
Jiménez, J. F. (2010). Una experiencia de desarrollo agrícola al sur de España: La localidad de El Ejido (An experience of agricultural development in the south of Spain: The town of El Ejido). In P. Sánchez Vera & A. Riella (Eds.), *Globalización y perspectivas de la integración regional* (pp.225–248). Universidad de Murcia.
Jones, T., Ram, M., & Villares-Varela, M. (2019). Diversity, economic development and new migrant entrepreneurs. *Urban Studies, 56*(5), 960–976.
Jones, T., & Ram, M. (2024). Migrant business: The rise of new migrant business in Britain and beyond. In G. Meardi (Ed.), *Research Handbook on Migration and Employment* (pp. 137–151). Elgar.
Kadıoğlu Polat, D. (2020). "Now the German comes": The ethnic effect of gentrification in Berlin. *Ethnicities*, 20(1), 155–176.
Kalra, V. S. (2000). *From textile mills to taxi ranks: Experiences of migration, labour and social change*. Ashgate.
Kesteloot, C., & Meert, H. (1999). Informal spaces: The geography of informal economic activities in Brussels. *International Journal of Urban and Regional Research, 23*(3), 232–251.
King, R., Lulle, A., Conti, F., & Mueller, D. (2016). Eurocity London: A qualitative comparison of graduate migration from Germany, Italy and Latvia. *Comparative Migration Studies, 4*(3), https://doi.org/10.1186/s40878-016-0023-1
Kloosterman, R., van der Leun, J., & Rath, J. (1999). Mixed embeddedness: (In)formal economic activities and immigrant businesses in the Netherlands. *International Journal of Urban and Regional Research, 23*(2), 252–266.
Kloosterman, R., & Rath, J. (2001). Immigrant entrepreneurs in advanced economies: Mixed embeddedness further explored. *Journal of Ethnic and Migration Studies, 27*(2), 189–201.
Kloosterman, R., & Rath, J. (2018). Mixed embeddedness revisited: A conclusion to the symposium. *Sociologica, 12*(2), 103–114.
Kofman, E., & Raghuram, P. (2015). *Gendered migrations and global social reproduction*. Palgrave Macmillan.
Korac, M. (2013). Transnational pathways to integration: Chinese traders in Serbia. *Sociologija, 15*(2), 245–260.
Lan, T. (2015). Industrial district and the multiplication of labour: The Chinese apparel industry in Prato, Italy. *Antipode, 47*(1), 158–178.
Laslett, B., & Brenner, J. (1989). Gender and social reproduction. *Annual Review of Sociology, 15*(1), 381–404.

Leontidou, L. (1990). *The Mediterranean city in transition: Social change and urban development*. Cambridge University Press.

Lindio-McGovern, L. (2012). *Globalization, labor export and resistance: A study of Filipino migrant domestic workers in global cities*. Routledge.

Lutz, H. (Ed.). (2008). *Migration and domestic work: A European perspective on a global theme*. Ashgate.

Lutz, H. (2011). *The new maids: Transnational women and the care economy*. Zed Books.

Marchetti, S. (2022). *Migration and domestic work*. Springer.

Marouda, M. (2020). A tale of three marketplaces: Chinese commodities, European fairs, Vietnamese entrepreneurs. *Global Networks, 20*(4), 725–745.

May, J., Wills, J., Datta, K., Evans, Y., Herbert, J., & McIlwaine, C. (2007). Keeping London working: Global cities, the British state and London's new migrant division of labour. *Transactions of the Institute of British Geographers, 32*(2), 151–167.

McDowell, L. (2013). *Working lives: Gender, migration and employment in Britain, 1945–2007*. Wiley-Blackwell.

McDowell, L., & Dyson, J. (2011). The other side of the knowledge economy: "Reproductive" employment and affective labours in Oxford. *Environment and Planning A: Economy and Space, 43*(9), 2186–2201.

Meier, L. (Ed.). (2014). *Migrant professionals in the city: Local encounters, identities and inequalities*. Routledge.

Milutinovic, S. (2008). Chinese transnational entrepreneurs in Budapest and Belgrade: Seeking markets, carrying globalization. In M. P. Smith & J. Eade (Eds.), *Transnational ties: Cities, migrations, and identities* (pp. 91–106). Transaction Publishers.

Näre, L. (2011). The moral economy of domestic and care labour: Migrant workers in Naples, Italy. *Sociology, 45*(3), 396–412.

Nyíri, P. (2002). Chinese in Hungary and their significant others: A multi-sited approach to transnational practice and discourse. *Identities, 9*(1), 69–86.

Parreñas, R. S. (2000). Migrant Filipina domestic workers and the international division of reproductive labor. *Gender and Society, 14*(4), 560–580.

Parreñas, R. S. (2012). The reproductive labour of migrant workers. *Global Networks, 12*(2), 269–275.

Parzer, M., & Huber, F. J. (2015). Migrant businesses and the symbolic transformation of urban neighbourhoods: Towards a research agenda. *International Journal of Urban and Regional Research, 39*(6), 1270–1278.

Peake, L., Koleth, E., Tanyildiz, G. S., Reddy, R. N., & Patrick, D. (2021). *A feminist urban theory for our time: Rethinking social reproduction and the urban*. Wiley.

Pérez, I., & Stallaert, C. (2016). The professionalization of paid domestic work and its limits: Experiences of Latin American migrants in Brussels. *European Journal of Women's Studies, 23*(2), 155–168.

Però, D. (2020). Indie unions, organizing and labour renewal: Learning from precarious migrant workers. *Work, Employment and Society, 34*(5), 900–918.

Plöger, J., & Becker, A. (2015). Social networks and local incorporation: Grounding high-skilled migrants in two German cities. *Journal of Ethnic and Migration Studies, 41*(10), 1517–1535.

Polyzou, I. (2020). Invest the city by doing businesses: Geographies of migrant entrepreneurs in the center of Athens. *Greek Review of Social Research, 153*, 61–81.

Portes, A. (2000). Immigration and the metropolis: Reflections on urban history. *Journal of International Migration and Integration, 1*(2), 153–175.

Portes, A., Castells, M., & Benton, L. A. (1989). *The informal economy: Studies in advanced and less developed countries*. John Hopkins Press.

Raco, M., & Kesten, J. (2018). The politicisation of diversity planning in a global city: Lessons from London. *Urban Studies, 55*(4), 891–916.

Rath, J. (2000). A game of ethnic musical chairs? Immigrant businesses and the formation and succession of niches in the Amsterdam economy. In S. Body-Gendrot & M. Martiniello (Eds.),

Minorities in European cities: The dynamics of social integration and social exclusion at the neighbourhood level (pp. 26–43). Macmillan.
Räuchle, C., & Schmiz, A. (2019). Migrant economies: Opportunity structures and potential in different city types. *Ethic and Racial Studies, 42*(1), 1766–1787.
Reyneri, E. (1998). The role of the underground economy in irregular migration to Italy: Cause or effect? *Journal of Ethnic and Migration Studies, 24*(2), 313–331.
Rigo, E. (2022). *La Straniera: migrazioni, asilo, sfruttamento in una prospettiva di genere* (The Strangeress: Migration, asylum and exploitation from a gender perspective). Carocci.
Rodgers, P. S., Shahid, M., & Williams, C. W. (2019). Reconceptualizing informal work practices: Some observations from an ethnic community in urban UK. *International Journal of Urban and Regional Research, 43*(3), 476–496.
Roy, A. (2005). Urban informality: Toward an epistemology of planning. *Journal of the American Planning Association, 71*(2), 147–158.
Samers, M. (2002). Immigration and the Global City Hypothesis: Towards an alternative research agenda. *International Journal of Urban and Regional Research, 26*(2), 389–402.
Samers, M. (2004). The "underground economy", immigration and economic development in the European Union: An agnostic-skeptic perspective. *International Journal of Economic Development, 6*(3), 199–272.
Sassen, S. (1991). *The global city. New York, London, Tokyo.* Princeton University Press.
Sassen, S. (1996). New employment regimes in cities: The impact on immigrant workers. *Journal of Ethnic and Migration Studies, 22*(4), 579–594.
Schmoll, C., & Semi, G. (2013). Shadow circuits: Urban spaces and mobilities across the Mediterranean. *Identities, 20*(4), 377–392.
Shaw, S., Bagwell, S., & Karmowska, J. (2004). Ethnoscapes as spectacle: Reimaging multicultural districts as new destinations for leisure and consumption. *Urban Studies, 41*(10), 1983–2000.
Sik, E., & Wallace, C. (1999). The development of open-air markets in east-central Europe. *International Journal of Urban and Regional Research, 23*(4), 697–714.
Smith, A., Stenning, A., Rochovská, A., & Świątek, D. (2008). The emergence of a working poor: Labour markets, neoliberalisation and diverse economies in post-socialist cities. *Antipode, 40*(2), 283–311.
Stenning, A., & Dawley, S. (2009). Poles to Newcastle: Grounding new migrant flows in peripheral regions. *European Urban and Regional Studies, 16*(3), 273–294.
Stock, M., & Schmiz, A. (2019). Catering authenticities. Ethnic food entrepreneurs as agents in Berlin's gentrification. *City, Culture and Society, 18*, 1–8.
Turaeva, R. (2013). Post-Soviet uncertainties: Micro-orders of central Asian migrants in Russia. *Inner Asia, 15*(2), 273–292.
Vaiou, D. (2002). In the interstices of the city: Albanian women in Athens. *Espace Populations Sociétés, 3*, 373–385.
van Doorn, N., & Vijay, D. (2024). Gig work as migrant work: The platformization of migration infrastructure. *Environment and Planning A: Economy and Space, 56*(4), 1129–1149.
van Doorn, N., Ferrari, F., & Graham, M. (2023). Migration and migrant labour in the gig economy: An intervention. *Work, Employment and Society, 37*(4), 1099–1111.
van Hooren, F. (2012). Varieties of migrant care work: Comparing patterns of migrant labour in social care. *Journal of European Social Policy, 22*(2), 133–147.
Villares-Varela, M., Ram, M., & Jones, T. (2018). Bricolage as survival, growth and transformation: The role of patch-working in the social agency of migrant entrepreneurs. *Work, Employment and Society, 32*(5), 942–962.
Ward, G. (1977). *Fort Grunwick*. Temple Smith.
White, P., & Hurdley, L. (2003). International migration and the housing market: Japanese Corporate movers in London. *Urban Studies, 40*(4), 687–706.
Williams, C. C. (2023). *A modern guide to the informal economy*. Elgar.
Wills, J., Datta, K., Evans, Y., Herbert, J., May, J., & McIlwaine, C. (2010). *Global cities at work: New migrant divisions of labour*. Pluto Press.

Wilpert, C. (1998). Migration and informal work in the New Berlin: New forms of work or new sources of labour. *Journal of Ethnic and Migration Studies, 24*(2), 269–294.
Wilson, K., & Portes, A. (1980). Immigrant enclaves: An analysis of the labor market experiences of Cubans in Miami. *American Journal of Sociology, 86*(2), 295–319.
Yanasmayan, Z. (2014). Further stay or return? Insights from the highly educated Turkish migrants in Amsterdam, Barcelona and London. In L. Meier (Ed.), *Migrant professionals in the city: Local encounters, identities and inequalities* (pp. 21–39). Routledge.

Chapter 4
Arrival, Settlement and the Remaking of European Urban Life

4.1 Introduction

Many cities in Europe possess a complex patchwork of places where newcomers arrive, settle and live out their daily lives. London is a prime example. In 2005, *The Guardian* newspaper published a map of 100 sites in the British capital associated with different national, ethnic and religious groups as the centrepiece of its special edition 'London: The World in One City'.[1] The map was presented to readers as the culmination of an in-depth investigation by a journalist who had 'spent months travelling across the capital, locating and visiting [...] immigrant communities' (Benedictus, 2005). As an accessible guide to the city's multicultural geography, it confirmed claims circulating in the media at the time that London had become the 'most diverse city ever' (ibid).

Anyone familiar with London would not be surprised by some of the connections identified between groups and places, such as Bangladeshis on Brick Lane, Punjabi Sikhs and Pakistanis in the western suburb of Southall or Jamaicans in Brixton. But the map also highlighted lesser-known 'communities', such as Koreans in South Malden, Portuguese in Stockwell and Algerians in Finsbury Park, some of which were a direct result of the recent increase in global migration to the city. The map appeared to capture a city seemingly at ease with its diversity. The celebratory tone also resonated with local government discourses about London as a global crossroads, which have prevailed in various guises over the past two decades, regardless of changes in political leadership (Raco & Kesten, 2018).

In September 2019, Catherine Blaiklock, founder of the Brexit Party (renamed Reform UK in 2021), offered a pithier portrait of multicultural London when she tweeted the following description of the city:

[1] The map and accompanying articles can be accessed here: https://www.theguardian.com/uk-news/series/london.

N. Dines, *Migration and European Cities*, IMISCOE Research Series,
https://doi.org/10.1007/978-3-032-15379-1_4

> Clapham looks like the Caribbean. Tower Hamlets looks like Pakistan or Syria. Mornington Crescent looks like Turkey. Redbridge looks like a foreign country. And Bermondsey and Stratford just look like one big non-English mess.

Here, the city's diversity was implicated in claims of excess and loss, while the hyperbolic claims typified the negative representations of immigration, past and present, that had accompanied the debate on Brexit. Blaiklock later deleted her tweet, but not before it was screenshotted by the campaign group Justice or Else UK and posted on their Facebook page.[2] Some of the subsequent posts echoed the group's accusations of racism, but most pilloried Blaiklock for her lack of knowledge of London. One commenter exclaimed, 'Clapham??? Last time I checked that place had more white people than white walkers'; to which another retorted: 'It's almost like she's using clichés from 20 years ago and hasn't visited any areas since, if you want white hipsters willing to spend £15 on avocado toast you go to Clapham.'

Despite their sharply contrasting visions of London, *The Guardian's* map and Blaiklock's tweet actually share similar features. Both indulge in the common practice of articulating the migration-city nexus through the identification of places where newcomers are seen to live and congregate. Both give the impression that migrants and people with migration backgrounds *stick together*, both measure this presence against an unspoken host society and both assert group unity on the basis of presumed national origin, ethnicity or racialized identities. Juxtaposed, they help us think critically about routine representations of migration-driven diversity in cities.

First, these acts of labelling difference raise questions about who gets to define the city and the impact that their definitions might have on public knowledge. It is one thing for the descriptions of places by liberal cosmopolitans or anti-immigrant populists to set the boundaries of public debate. It is quite another when the perspective is that of a recently arrived asylum seeker or a settled resident with a migration background whose engagement with multiculturalism may be 'ordinary, banal, unremarkable and sometimes mundane to the point of boredom' (Back et al., 2018: 137). In such cases, opportunities to influence dominant perceptions of urban diversity are usually more limited.

Second, identifying the traces of migration in urban contexts requires not only accuracy but also an awareness of the complexity of urban areas and demographic groups. Hence, the task is not just to point out the blunders made by someone like Blaiklock, but also to recognize the implications of 'correctly' associating a place with one or more circumscribed groups of people. Scholars have long highlighted the limitations of reading migratory processes through an ethnic lens (Glick Schiller et al., 2006), and the importance of paying attention to a range of variables that constitute social and cultural difference in a given place, and the connections that people cultivate beyond the immediate confines of their neighbourhood and city (Tasan-Kok

---

[2] Justice Or Else UK describes itself as 'a black-led radical direct action movement [that] fight[s] against black oppression, racism and deaths in police custody' (https://www.facebook.com/pg/JusticeOrElseUK/about/). The Facebook post on Catherine Blaiklock's tweet is dated 29 September 2019. The original tweet can be found here: https://web.archive.org/web/20190928190830/https:/twitter.com/blaiklockBP/status/1177877514254999552.

et al., 2014; Vertovec, 2007). Nevertheless, as the anthropologist Gerd Baumann illustrated in his study of Southall, ideas about ethnicity and community often remain important reference points for members of migrant and minority groups. Baumann distinguishes between *dominant* discourses of ethnicity, used by institutions and the media, which tend to reify culture and community, but which are also appropriated by minority groups to stake their claim to an area and compete for resources, and *demotic* (or everyday) discourses, which allow for the continual creation of new ideas of community that transcend established 'ethnic' divisions (Baumann, 1996: 10).

Finally, descriptions of migrant spaces, regardless of their attention to detail, tend to produce a compartmentalized picture of the city that eliminates from view a host of other processes that reconfigure urban life. This is illustrated in the above quip about Clapham: the issue is not just that Blaiklock makes false and racist aspersions about this part of south London, but that her ignorance of gentrification in the same area makes her claim of 'ethnic replacement' sound ludicrous.

In other words, while we might reasonably argue that locating migrant groups in the city is an inescapable way of making sense of urban diversity, on closer inspection the practice raises more questions than it answers. Why do newcomers end up where they are, and how might they be able to move on? Why, as both *The Guardian* and Catherine Blaiklock suggest, do some group members concentrate in certain places and not others? What choices are available to different migrants at any given moment, and how might these be constrained by factors beyond their control? How do histories of settlement influence contemporary residential patterns? How do migrants use and remake the shared spaces of the city, and how do they carve out more private, intimate worlds?

Reflecting on the ways in which urban diversity is labelled opens up a set of issues that are the focus of this chapter. The first section looks at migrants' access to housing in cities and related academic debates over the last few decades. The second, and main, part of the chapter addresses the broader theme of settlement, unpacking some key methodological dilemmas in navigating the literature and using the case of Bangladeshi migration to London and Rome to underline the divergent paths and outcomes of migrant settlement in different urban contexts. It then considers the extensive literature on segregation in different European cities and how this phenomenon has often been politicized in public debate, before turning to the burgeoning body of largely qualitative research on residential mobility, transnationalism and migrant homemaking. The third section focuses on what sociologist Suzanne Hall (2015) has termed 'migrant urbanisms', in other words the spaces and infrastructures that migrants use and create to sustain urban livelihoods, while the final section considers everyday social interactions at the urban scale in relation to discussions about conviviality, racism and violence.

This chapter draws on a range of literature from across the social sciences, including urban geography, sociology, anthropology and planning. Over the last fifty years, the vast majority of research and debate has focused on the diverse settlement experiences of migrants from non-Western or lower-income countries and, to varying degrees, their interconnections with broader social and spatial transformations of European cities. This particular focus is reflected in the present chapter.

More recently, a growing body of research has investigated the housing and residential trajectories of the increasing number of highly skilled migrants and 'middling transnationals'[3] from both the global North and South. Where relevant, this literature is referenced to draw further attention to the complexity of migrant settlement processes.

4.2 Arrival: Housing Opportunities and Barriers

The type and location of housing available to newcomers depends, to some extent, on variables operating at the individual and group level, such as a person's financial resources and social networks, the timing and circumstances of their migration and their ability to pre-arrange accommodation before arrival (Robinson et al., 2007; Wessendorf, 2018). Options are also to a certain degree shaped by people's pre-existing housing preferences and aspirations, their knowledge of destination cities and fluency in a local language (Maslova & King, 2020). At the same time, structural and systemic factors play a crucial role in shaping the opportunities in a given city. These include a city's housing system, i.e. how provision and consumption are organized (Arbaci, 2019); planning, welfare and migration policies that affect, among other things, access to social housing (Dotsey & Chiodelli, 2021); levels of racism and discrimination in housing markets (McKee et al., 2021); and the housing choices of 'native' residents (Andersen, 2017). In addition, scholars have recently turned their attention to actors who intermediate between supply and demand, such as brokers who provide housing services to migrants who would not otherwise have access to certain markets (Bernt et al., 2022).

Housing opportunities vary between urban areas, over time and for different groups of migrants. During the post-war period of mass labour migration, guest workers in northern and central European cities often had no choice but to live in communal facilities provided by the state or employers. Options for non-recruited migrant workers and their families were limited to cheap hotels, boarding houses or the private rented sector, where discriminatory practices by banks and estate agents blocked access to mortgages and better housing stock, leading many to accept substandard and overcrowded accommodation (Castles & Kosack, 1973). Once an individual or group had accumulated sufficient capital, they bought cheap properties, usually in less desirable inner-city areas, which, if large enough, were often subdivided and rented out to other group members.

Until the 1970s, the social housing sector, which had expanded massively in northern and western Europe after the Second World War, was largely inaccessible to foreign-born residents, either because they were excluded as non-citizens or because

[3] In recent years, a number of scholars have used the term 'middling-transnationals' to refer to middle-class skilled migrants. This builds on the original idea of geographers David Conradson and Alan Latham, who proposed studying 'middling' forms of transnational mobility as a means of countering the 'narratives of trouble-free movement of disembedded [elites]' that had previously dominated transnational research (Conradson & Latham, 2005: 227).

public institutions and housing agencies prioritized the local population. John Rex and Robert Moore's landmark study of housing and race relations in the Sparkbrook district of Birmingham, UK, found that there was a de facto 'colour bar' for Asian and Caribbean residents due to the organization of the waiting list, which required at least five years residence and gave extra points for war service and length of time on the list (Rex & Moore, 1967). Immigrants in Swedish cities appeared to fare better than elsewhere due to an ambitious housing policy manifested in the so-called 'Million Dwellings Programme' between 1965 and 1974, which meant that many moved into newer and higher quality segments of the housing stock, albeit often on the outskirts of cities (Murdie & Borgegård, 1998).

The housing options for people arriving in European cities today appear to be markedly different. First, housing costs have risen sharply in many cities, especially over the last two decades. The financialization of housing, overheated urban property markets, the privatization and depletion of social housing stocks and the end of rent controls have combined to fuel a crisis in the provision of decent, affordable housing across Europe and beyond, which has hit recent low-wage migrants particularly hard (Lombard, 2023). The poor and undesirable neighbourhoods that once existed in and around the city centres of north-western Europe are largely a thing of the past, although many such areas still provide crucial 'arrival infrastructures' in the form of support networks, casual work opportunities and temporary accommodation that enable migrants to meet their immediate housing needs and to gain a foothold in a city before moving on (Meeus et al., 2019). Opportunities in expensive global cities like London have become more spatially dispersed, with some newcomers commuting from distant, lower-priced suburbs or bypassing large metropolises altogether in favour of smaller, cheaper and increasingly multicultural urban centres such as Milton Keynes, located 80 kms northwest of the UK capital (Neal et al., 2013).

Second, the diversification of international migration since the 1990s has created a more complex variety of housing experiences within and between European cities. For example, while the options for many migrants in major global and tourist cities are constrained by prohibitive costs, a minority of transnational professional elites and lifestyle migrants have successfully competed in property markets, sometimes creating new submarkets that contribute to exacerbating housing crises for locals and migrants alike (Cocola-Gant & Lopez-Gay, 2020; White & Hurdley, 2003). By contrast, many European skilled migrants in these same cities are far less flexible in their initial housing choices. Due to the shortage of affordable housing and intense competition in the private rented sector, the early stages of their housing careers often involve a trade-off between the high cost of a desirable location or comfortable accommodation, and reducing costs by sharing (Manting et al., 2024; Maslova & King, 2020). Meanwhile, the increases in asylum-seeker flows to Europe since the early 2010s, coupled with national dispersal policies, has led to the emergence of urban reception systems, mainly in large-capacity accommodation such as requisitioned hotels, disused barracks and purpose-built camps (Kreichauf, 2018). These systems are designed to provide temporary shelter, although in some cases people end up staying for several months or years, and on exit former residents typically face high levels of housing insecurity or, where such provision exists, are channelled into the least desirable social housing in cities (Bernt et al., 2022).

Third, increasingly restrictive national immigration policies have had a direct negative impact on housing opportunities. Studies have highlighted how the tightening or removal of access to housing on the basis of legal status is a key cause of precarious housing conditions (Dietrich-Ragon, 2017; Dotsey & Ambrosini, 2025). Informal rental arrangements, which are particularly common in large southern European cities, pose serious problems for obtaining residency, which is necessary to access local welfare and renew legal documents. Another important development is the formal involvement of housing authorities and private landlords in immigration management. For example, the 'right to rent' provision of the 2016 UK Immigration Act, which makes it a criminal offence to provide accommodation to undocumented migrants, turns housing providers into de facto border agents (McKee et al., 2021).

Nevertheless, significant continuities with the past remain. As Deborah Phillips has observed, almost half a century after the publication of Rex and Moore's Sparkbrook study, 'newness, outsider status (legally and socially constructed), poverty, fear of harassment and lack of entitlements still shape newcomers' life chances, housing experiences, personal security and locational options' (Phillips, 2015: 394). Although shifts in public sentiment, grassroots collective action and anti-discrimination legislation since the 1960s have mitigated the most extreme discriminatory practices in housing and mortgage markets, racism and anti-immigrant attitudes continue to be entrenched in the unequal distribution of resources and opportunities (Verhaeghe & De Coninck, 2022). For the majority of newcomers who are not allocated accommodation on arrival or who have low incomes, housing pathways remain heavily dependent on social networks and the private rented sector. Situations common to northern Europe in the 1960s—substandard housing, unscrupulous landlords and insecure tenancies—can now be found in cities across southern and eastern Europe, where international migration is more recent, social housing stock and tenancy rights are more limited, and native levels of home ownership are higher than in the 'old' destination countries (Dotsey & Chiodelli, 2021). Meanwhile, competition for increasingly scarce social housing continues to be a flashpoint for both anti-immigrant protests and cross-community campaigns for tenant protection and improved services (Hamann & Türkmen, 2020).

The lack of access to any kind of housing has sometimes been compensated by informal practices and direct action. In the post-war period, self-built settlements provided shelter for tens of thousands of internal labour migrants on the outskirts of southern European cities and for similar numbers of Algerian and Portuguese workers around Paris (Manzano Gómez, 2024), while the rise of migration from Eastern Europe in the 1990s led to the reappearance of smaller shantytowns in and around Rome and Paris (Mudu & Chattopadhyay, 2017). Informal practices are no longer exclusively associated with Mediterranean Europe, but have become a feature of northern cities, such as the 'beds in sheds' phenomenon in London where outbuildings in private gardens are illegally rented to mainly undocumented migrants (Schiller & Raco, 2021).

Another alternative housing strategy among migrants is the squatting of empty properties. These are often linked to political campaigns to secure better housing, as in the case of the Bengali Housing Action Group, which coordinated the mass occupation of semi-derelict buildings by Bangladeshi families in east London in the 1970s to pressure local authorities to rehouse them nearby (Glynn, 2010). Over the past two decades, migrant occupations have become a widespread phenomenon in Italian and Greek cities. In Rome, it is estimated that around 70–80% of the 12,000 people living in squatted buildings in 2018 were international migrants, including a growing number of refugees and asylum seekers who had crossed the Mediterranean (Montagna & Grazioli, 2019).

Finally, since the 1990s, European metropolitan regions have witnessed the rapid development of an archipelago of administrative detention centres (Arbogast & Migreurop, 2016). Such facilities can be found in or near the centres of numerous European cities, including Turin, Marseille, Madrid, Basel and Bucharest. Other centres have been established in more secure premises on the edges of cities, often close to international airports to facilitate deportations, such as Harmondsworth near London Heathrow, Ponte Galleria near Rome Fiumicino and Trandum near Oslo Gardermoen. Thus, for the approximately 200,000 undocumented migrants and failed asylum seekers detained in Europe each year,[4] today's 'arrival city' functions simultaneously as a 'carceral city' and a point of forced departure. Nevertheless, some centres have been permanently or temporarily closed or converted into reception accommodation, often as a result of changing immigration policy priorities, but sometimes in response to deaths and revolts of detainees and anti-detention campaigns by migrant rights activists (Spena, 2016).

4.3 Migrant Urban Settlement Processes

Since the time of the Chicago School, social scientists have sought to understand why urban newcomers end up living in certain parts of cities. Ernest Burgess famously identified immigrant 'ports of first entry' such as 'Little Sicily', 'Chinatown' and 'Black Belt' in the inner rings of Chicago, which drove the radial expansion of the city as residents gradually moved out to more attractive neighbourhoods in surrounding zones (Burgess, 1928: 106). From the 1960s onwards, scholars began to reject the Chicago School's ecological model of settlement, emphasising instead the impact of socio-economic and political constraints, and the choices that individuals or households make despite or because of these constraints. For example, John Rex and Robert Moore (1967) drew attention to the discriminatory practices of local authorities and estate agents that forced Commonwealth immigrants in 1960s Birmingham to live in the worst inner-city housing. The Pakistani anthropologist Badr Dahya (1974), who conducted fieldwork during the same period, challenged this view. He argued that

[4] This annual estimate is based on the aggregation of data collected by the Global Detention Project (https://www.globaldetentionproject.org/).

the concentration of Pakistanis in cheap terraced housing was the result of a rational choice because cohabitation suited their common religious and dietary requirements as well as their immediate goal of sending remittances home. In turn, Dahya, was criticized for conflating nationality with ethnicity and ignoring the internal class, cultural and linguistic differences within the Pakistani population (Saifullah Khan, 1976). Rather, the migrants' broadly similar lifestyle choices, coupled with local employment and housing conditions, fostered the construction of a sense of collective identity and solidarity. This reduced the need for contact with the 'majority society' and ultimately led to similar settlement trajectories (ibid: 222).

The interplay between choice and constraint remains a key framework for interpreting migrant settlement processes (McGarrigle, 2016; Peach, 1998). In parallel, a distinctly European research agenda has emerged since the late 1990s, particularly in the field of segregation studies. Breaking with the US-centric and globalization-driven metaphors of the 'dual city' and 'social polarization' that dominated Western urban theory in the late twentieth century, this approach has sought to account for the complex diversity of migration and residential patterns in European cities, bringing to the centre of analysis, among other things, the transformation of state-market relations and the role of different welfare regimes (Arbaci, 2019: 21–62).

4.3.1 Some Preliminary Remarks on Migrant Settlement Literature

In order to navigate through the extensive literature on migrant settlement in European cities, it is helpful to keep in mind three crucial elements: first, the aspect of settlement under analysis; second, the categorization and significance given to migration; and third, the attention paid to the urban context of research.

First, settlement is a multidimensional process and cannot be reduced to a single event or outcome. Whether settlement is seen as complete or ongoing depends on the perspective and criteria through which it is interpreted. It is therefore necessary to take into account the specific stages of settlement under consideration, the spatial and temporal scope of analysis, and the methods and data used. Research may examine residential patterns at a specific point in time or how they change over a longer period. It may focus on a particular city, neighbourhood, street or even a single building. It may use large data sets to map the distribution of groups or rely on qualitative fieldwork to capture contextual nuances. Attention also needs to be paid to the evaluative connotations associated with different processes and how these influence findings and analysis. This is important given the public controversy surrounding certain aspects of migrant settlement, as discussed below.

Second, the importance of migration in the literature on settlement processes can vary considerably. In some cases, the migration context is a central analytical focus; in others, it seems less relevant, either relegated to an event that began and ended in the past, or subsumed under the rubric of ethnicity, nationality or religious affiliation.

In part, this reflects national differences in the study of the phenomenon. But it also illustrates more generally the contradictions inherent in the idea of 'migrant settlement' itself (at what point during 'settlement' does one cease to be a 'migrant'?) and in the categorizations that constitute and maintain migrants as objects of analysis. Indeed, key sources of statistical information on migrant settlement, such as census data and housing surveys, are organized according to ethnicity or citizenship, and usually do not differentiate between individuals who have migrated and those born locally. Reliance on categories based on nationality and ethnicity can therefore obscure the heterogeneous social fabric of the city. However, these same categories have been crucial in conducting large-scale quantitative studies that shed light on socio-economic inequalities and systemic racism in housing markets. Moreover, while official statistics can be instrumentally interpreted to support popular stereotypes about the residential characteristics of different groups ('group x sticks together', 'group y lives in social housing', etc.), they are just as often mobilized to refute public claims that, for example, a particular minority is 'taking over' a neighbourhood (Finney & Simpson, 2009). In other words, when reviewing the literature, it is important to consider whether and how data on ethnicity and nationality are used and how they relate to the conceptualization of migration as a factor in settlement processes.

The final important element to consider is the extent to which researchers engage with the complexity of the local context. Over the past two decades, the literature on migrant settlement has examined an increasing range of European cities, shedding light on different patterns and processes and challenging the analytical reach of general theoretical models. In an early attempt to systematically explain variations in migrant settlement in European cities, the British geographer Paul White (1999) proposed a multi-level 'contextual structural model'. In addition to general factors operating at the global, regional and national levels, such as economic globalization, population decline and EU integration, White identified four major local forces that should be understood as influencing residential outcomes in individual cities: (1) the 'economic conjuncture', which refers to the particular economic opportunities available to people on and after arrival in a city; (2) the 'ideologies of the host society', i.e. the local societal attitudes and policies towards migrants and minority groups; (3) 'migrant and minority characteristics', which include the history of arrival and settlement, age structure, gender balance and average household size of a particular group with which an individual may be associated; and (4) 'urban context', i.e. the existing structure of a city and its particular housing market and planning regime (ibid: 163–172).

4.3.2 Settlement Pathways in Two Cities: Bangladeshis in London and Rome

To illustrate how the local forces identified by Paul White play out in different cities, I will briefly compare two principal centres of Bangladeshi settlement in Europe: Spitalfields in the London Borough of Tower Hamlets and Torpignattara in the Fifth Municipality of Rome.[5] At a first glance, these two neighbourhoods appear to have a lot in common: both are dense, traditionally low-income districts to the east of their respective city centres, and both have long histories of international and internal migration. Both neighbourhoods experienced post-war depopulation as former residents moved to outlying areas, both have a strong presence of 'ethnic' and particularly Bangladeshi businesses (both are sometimes referred to by the same English toponym 'Banglatown'), and in recent years both have felt the pressures of gentrification to varying degrees. In both cities, early Bangladeshi migration was predominantly male, with women tending to arrive later through family reunification.

Beyond these immediate similarities, however, the two neighbourhoods have distinct settlement patterns that reflect particular urban, economic, political and migration contexts. These include the different geographical origins of Bangladeshis, their different periods of arrival, and the different economic opportunities and housing markets in the two cities.

In London, significant numbers of Bangladeshis, mainly from the rural region of Sylhet, moved to Spitalfields in the 1960s and 1970s, attracted by the presence of a small Bengali community that had existed since the settlement of Lascars in the interwar years, by the area's proximity to sources of employment in the local garment industry and, later, in the curry restaurants that opened along the main thoroughfare, Brick Lane, and by the availability of cheap, if run-down, accommodation in the private rented sector and the prospect of access to the borough's extensive social housing stock. Indeed, since the 1980s, a large proportion of Bangladeshi households have been housed locally by Tower Hamlets Council, partly as a result of a campaign to house Bangladeshis close to community facilities to protect them from racist attacks, but also as a result of discriminatory allocation practices within the council that channelled households into the worst housing in the borough, which happened to be in and around Spitalfields (Alexander, 2011; Glynn, 2010).

In contrast, Bangladeshi migration to Rome began to increase significantly after Italy's first major immigration amnesty in 1990, and in the early years consisted mainly of university-educated men from Dhaka and the surrounding region, who often arrived in Italy via other locations in Europe. Many spent their first years

[5] London is made up of 32 local authorities known as boroughs. Tower Hamlets, which includes the inner parts of the city's East End and the redeveloped Docklands area, is London's tenth most populous borough and experienced a population growth of 22.1 per cent between 2011 and 2021, the highest rate in London and England. Rome is divided into 15 sub-administrative districts known as *municipi* (or municipalities). The Fifth Municipality, which is east of the historic centre and includes the 'historical peripheries' of Pigneto and Centocelle, is the third most populous in the city.

in precarious housing in the Esquilino district near the city's main railway station, either in extremely overcrowded apartments or in squatted buildings (Knights & King, 1998). Only after gaining a greater degree of financial and legal stability did some of this population move to Torpignattara, an area with a considerable supply of cheap and small properties to rent or buy on the private market (in contrast to Spitalfields, social housing in Torpignattara accounts for only 0.3% of the area's total stock), even if this meant commuting to other parts of Rome for low-paid jobs in catering and tourism (Pompeo, 2011). Over the next twenty years, the development of community facilities such as grocery stores and specialist travel agencies, and the increase in Bangladeshi chain migration to Rome, would turn Torpignattara into a gateway district.

The different settlement processes are reflected in the demographic profiles of the two neighbourhoods and the residential distribution of Bangladeshis in the two cities (Figs. 4.1 and 4.2). According to the 2021 UK Census, there were over 107,000 Bangladeshis in Tower Hamlets, accounting for 34.6% of the total population (up from 32% in the 2011 Census), making them the borough's largest 'ethnic' group. Most of this population is concentrated in the western and central wards including Spitalfields & Banglatown, where people of declared Bangladeshi origin make up over 41.6% of the population (slightly higher than the proportion recorded in the 2011 Census, but down from 58.1% in the 2001 Census). With a third of London's Bangladeshis living in Tower Hamlets, no other major minority group is so strongly represented in a single London borough, although the population has become more dispersed across neighbouring boroughs since the 1990s.

Not unlike Tower Hamlets, 29% of Rome's Bangladeshi population in 2021 lived in the fifth municipality, with 10% concentrated in Torpignattara. However, Bangladeshis make up only 4 and 7% of the total population of these two areas.

Ethnicity	**Spitalfields & Banglatown Ward** [population | % of total population]		**Tower Hamlets Borough**		**London**	
White British	2,208	16.69	71,177	22.94	3,239,281	36.81
White Irish	155	1.17	3,567	1.15	156,333	1.78
White Other	2,325	17.58	45,187	14.56	1,290,838	14.67
Indian	392	2.96	10,135	3.27	656,272	7.46
Pakistani	164	1.24	3,341	1.08	290,549	3.30
Bangladeshi	**5,501**	**41.59**	**107,333**	**34.59**	**322,054**	**3.66**
Black African	444	3.36	15,373	4.95	697,054	7.92
Black Caribbean	107	0.81	4,930	1.59	345,405	3.93
Mixed	543	4.11	15,409	4.96	505,775	5.75
Other	1,387	10.49	33,854	10.91	1,296,164	14.72
Total	13,226	100.0	310,306	100.0	8,799,725	100.0

Fig. 4.1 Population breakdown by declared ethnicity in Spitalfields and Banglatown, Tower Hamlets and London (2021). *Source* 2021 UK Census, https://www.ons.gov.uk & https://data.london.gov.uk/dataset/2021-census-wards-ethnicity-language-identity-religion

Citizenship	Torpignattara Urban Zone [population | % of total population]		Fifth Municipality		Rome	
Italy	35,932	77.07	196,303	82.31	2,434,483	86.53
Romania	968	2.08	6,998	2.94	86,665	3.08
Philippines	783	1.68	2,734	1.15	41,049	1.46
Bangladesh	**3,320**	**7.12**	**9,679**	**4.06**	**33,367**	**1.19**
China	1,961	4.21	5,056	2.12	19,206	0.68
Ukraine	223	0.48	1,211	0.51	14,938	0.53
Peru	243	0.52	1,250	0.52	12,546	0.45
India	285	0.61	1,214	0.51	11,680	0.41
Egypt	360	0.77	1,653	0.69	11,197	0.40
All other groups	2,549	5.46	12,389	5.19	148,234	5.27
Total	46,624	100.0	238,487	100.0	2,813,365	100.0

Fig. 4.2 Population breakdown by national citizenship in Torpignattara, Fifth Municipality and Rome (2021). *Source* Popolazione iscritta in anagrafe—stranieri, Roma Capitale, https://www.comune.roma.it/web/it/dettaglio.page?contentId=PAG23000

Moreover, they represent less than a third of the total foreign population of Torpignattara, which, like most 'migrant neighbourhoods' in Rome and other Italian cities, is characterized by a high mix of different national groups. In sum, the common denominator of 'Banglatown' masks two neighbourhoods with quite different settlement histories and shifts in the composition of the local Bangladeshi population over time.

4.3.3 Residential Segregation

The settlement patterns of Bangladeshis in London and Rome invoke a number of spatial metaphors and concepts that have long been the focus of scholarly attention: 'dispersal', 'clustering', 'segmentation' (which refers to the concentration of population groups in different housing sectors) and, last but not least, 'segregation'. This latter term refers to the spatial separation of a subgroup from the wider population and can be explained simply as follows: 'A group is said to be completely mixed in a spatial sense when its members are distributed uniformly throughout the population. The greater the deviation from a uniform dispersal, the greater the degree of segregation' (van Kempen & Özüekren, 1998: 1632). Thus, based on the data above, in 2021 Bangladeshis in London were more spatially segregated than those in Rome.

Segregation is seen as an important lens through which to assess migrant settlement processes. It can have potentially detrimental effects, for example on people's social and economic participation in local society, but it can also offer benefits, particularly for newcomers, by providing entrepreneurial opportunities and protection from discrimination. Segregation has indeed become a prominent subfield in the study of migration and cities, encompassing a variety of approaches, from quantitative analyses that calculate residential distribution—the simplest and most popular measure being the 'index of dissimilarity', which compares how one subgroup is geographically distributed in relation to another—to critical studies that examine and debate the causes and consequences of segregation (for an overview, see Musterd, 2023).

Cities have always been characterized by spatial segregation, not only on the basis of migration status, race or ethnicity but also along lines of income, class, age, gender and religion. The concept of segregation was first developed in response to the ethnic and racial divisions in US cities in the early twentieth century. In the 1990s, European scholars began to move away from the prevailing theories of North American studies, such as Sassen's social polarization thesis, to focus on how socio-economic structures and institutional mechanisms reproduce spatial inequalities and influence people's residential trajectories over time and across cities (Arbaci, 2019: 25–34). Segregation was understood as a multifactorial and context-dependent process, and was considered to be relatively low in Europe compared to the extreme levels in the United States (Musterd et al., 1998).

Most of the early research on the residential segregation of migrant and minority groups in Europe focused on northern European cities, particularly in Scandinavia, the Netherlands and Britain (Murdie & Borgegård, 1998; Peach, 1999; van Kempen & van Weesep, 1998). A pioneering comparative study of Amsterdam, Brussels, Frankfurt, Düsseldorf, Manchester, London, Stockholm and Paris interpreted different patterns of segregation in terms of local immigration histories, urban development processes, and the functioning of housing markets (Musterd et al., 1998), while an edited volume by the same authors analyzed the role of the welfare state (Musterd & Ostendorf, 1998). These early findings paved the way for a field of inquiry that has sought to describe and explain changing levels of spatial segregation, not only between different cities but also at different geographical scales (Catney, 2018) and across different domains, such as schooling (Rangvid, 2007) and the workplace (Marcińczak et al., 2015).

However, the residential patterns and contextual factors mapped in northern Europe—with its particular housing and welfare systems, experiences of deindustrialization, and post-war migration flows—have not been replicated in other parts of the continent. In Central and Eastern Europe, most studies have focused on the development of segregation of 'native' residents after the end of state socialism, including discrimination against Roma populations (Sýkora, 2009) and the continued concentration of post-1945 Russian and Ukrainian immigrants and their descendants in Soviet-built housing estates in the suburbs of Tallin, Vilnius and Riga (Tammaru et al., 2016: 287–357). Only in recent years have scholars begun to investigate the spatial distribution of international migrants (e.g. Stonawski et al., 2022 on Krakow).

Meanwhile, research in the Euro-Mediterranean region has pointed to the limited analytical purchase of northern European models of ethnic segregation (see Malheiros, 2002 on Lisbon; Leal, 2004 on Madrid; Maloutas, 2007 on Athens). Sonia Arbaci's (2019) comparative study of ethnic segregation in Portugal, Spain, Italy and Greece argues forcefully for the need to pay attention to a different set of contextual factors: family-based welfare systems, limited social housing provision, high private ownership, histories of informal housing, different labour markets and later international migration flows. Moreover, Arbaci observes an inverse correlation between spatial and social inequality: while migrant residents in Rome or Madrid face higher levels of social marginalization than in northern European cities such as Copenhagen or Stockholm with their more extensive welfare and social housing systems, they are actually less spatially segregated (ibid: 1–3). In other words, while parts of southern European cities tend to be more socially mixed and migrants are more spatially dispersed, this does not mean that the cities are more equal or that migrants are more integrated into local society.

Residential patterns in southern Europe also reflect particular socio-spatial structures and forms of urban development. For example, Athens and Naples are both characterized by vertical segregation in their city centres, with low-income migrants and mixed-income 'natives' living on different floors of the same buildings. The reasons for this peculiar phenomenon are different: in Athens it is the result of rapid post-war urbanization, fuelled by internal migration and the 'antiparochi' system, which incentivized private construction and led to different social classes occupying the same blocks; while in Naples it is rooted in the intense population growth of the early modern period, which saw the ground floors of tenement blocks converted into housing for the urban poor. In both cities, depopulation, suburbanization and the end of secure tenancies led to low quality but centrally located properties being rented to or purchased by migrant residents (Dines & Mattiucci, 2022; Maloutas & Spyrellis, 2016). These examples of vertical micro-segregation not only underline the complexity of migrant settlement patterns in European cities but also challenge conventional 'horizontal' approaches to mapping segregation.

Systematic comparative research on ethnic and migrant residential segregation in European cities has made a significant contribution to the field of urban studies. It has brought contextual diversity to the fore and has dispelled the idea of a single European urban theory. Furthermore, scholars have actively exposed the misuse of terminology and inaccurate claims made by public figures (Peach, 2005).

Nevertheless, it is worth noting the general criticisms that have been levelled at this body of work. First, macro approaches to ethnic segregation have tended to produce a somewhat static understanding of residential location in which the element of change is reduced to a stop-motion sequence dictated by the intervals of available data within fixed areal units. If the causal factors of segregation have typically been seen as complex, the complexity of segregation itself has often boiled down to the production of indices and choropleth maps, with little attention paid to its temporal and relational dimensions. Drawing on qualitative fieldwork as well as digital methods such as mobile phone data analysis, segregation scholars have sought to address these shortcomings by considering, among other things, people's spatial activities and networks within and beyond their residential areas (Piekut, 2021), and the unsynchronized nature of migration and life courses that influence the concentration and dispersion of groups over time (McGarrigle, 2016).

A second criticism is the lack of attention paid to issues of race and racism in European segregation studies (Picker, 2017). Indeed, it is commonly assumed that race is more relevant in the United States than in Europe, where residential segregation of minority groups is seen as more correlated with class divisions (Arbaci, 2019: 44). Researchers have redressed this absence by bringing European histories of racism and colonialism to bear on the analysis of segregation, not only in terms of the everyday processes of racialization that create boundaries between groups at the neighbourhood level (Gruner, 2010) and fuel discriminatory housing policies (Risager, 2023), but also as elements structurally embedded in the production of urban spatial inequalities. This latter aspect is exemplified by the marginal and substandard accommodation allocated by local authorities to indigenous and migrant Roma populations in European cities (Picker, 2017).

4.3.4 The Politicization of Segregation: No-Go Areas, Ghettos and Territorial Stigma

As a scientific concept, segregation is used by researchers to describe the unequal distribution of population groups and therefore, at least hypothetically, has no predetermined positive or negative connotations. In media and policy debates, however, segregation is often conflated with *ethnic residential* segregation and considered to be inherently problematic. Beyond the real or imagined consequences of segregation and the subtle and not-so-subtle prejudices that influence perceptions of the residential distribution of different groups, this negative portrayal has been exacerbated by a paradigm shift in housing and urban policy. Although quotas limiting the spatial concentration of immigrants in public housing have existed since the 1970s (Grillo, 1985: 125–128), economic and state restructuring in recent decades has resulted in a move away from comprehensive redistributive approaches to housing needs towards neoliberal area-based programmes that identify poor neighbourhoods as incubators of social ills (Arbaci, 2019: 44–46). The clustering of migrant households in deprived areas is thus interpreted as both a cause and a symptom of their lack of 'integration' (Finney & Simpson, 2009), despite the fact that there is no direct link between the spatial dimension of segregation and socially constructed notions of integration. Indeed, as studies of micro-segregation show, the physical proximity of different groups does not guarantee interaction (Gruner, 2010), while those groups usually considered to be well integrated into society, such as high-income elites, are also among the most segregated (Phillips, 2007).

In recent years, political discourses on segregation have largely targeted low-income urban areas with high levels of social or private rented housing and non-white and, in particular, Muslim groups. In 2024, a British Conservative MP for an affluent London suburb caused a minor media furore after describing Bangladeshi-dominated parts of Tower Hamlets as 'no-go' areas where other people were 'fearful of going out' (Jessup & Jackson, 2024). Such claims were by no means new, exceptional or confined to right-wing dog-whistle politics. Indeed, one of the most notorious commentaries on segregation in recent British history was a high-profile speech in

2005 entitled 'After 7/7: Sleepwalking into Segregation' delivered by Trevor Phillips, a former Labour Party politician, son of Guyanese immigrants and then Director of the Commission for Racial Equality (for detailed analyses of the speech, see Finney & Simpson, 2009; Slater, 2021). According to Phillips, the emergence of 'marooned communities', particularly among South Asian Muslims, was a root cause of the urban riots involving Pakistani and white youths in northern English cities in 2001 and the milieu that partly produced the London 7/7 bombers. Tolerance of diversity had enabled division and extremism. The call for common codes of behaviour and respect resonated with the neo-communitarian agenda of the New Labour government, with its emphasis on social cohesion and civic responsibility. The problem with Phillips's argument, as experts subsequently pointed out, was that the levels of segregation in the urban areas mentioned in the speech had actually been falling for some time (Finney & Simpson, 2009; Peach, 2009; Phillips, 2006).

The most controversial element of the 2005 speech was Phillips's use of the word 'ghetto' to emphasize the gravity of the situation. According to Phillips, 'some districts are on their way to becoming fully fledged ghettos—black holes into which no-ones goes without fear and trepidation, and from which no-ever escapes undamaged' (quoted in Finney & Simpson, 2009: 116). Historically, the word ghetto referred to the involuntary spatial confinement of a targeted group, as in the case of the first Jewish ghetto in sixteenth-century Venice. The modern version of the ghetto is indelibly associated with US cities and is understood to have two characteristics: first, a single ethnic or racial group constitutes almost the entire population of a residential district, and second, most members of that group in a city are located in the same district (Peach, 1996: 216–217). Philips's claim was roundly rejected by segregation scholars at the time on the grounds that no urban area in contemporary Britain or Europe possessed such characteristics (Peach, 2005; Wacquant, 2007). However, as many critics insisted, this was not simply a dispute about numbers. The main issue was Phillips's endorsement of a racialized vocabulary which has often had an instrumental role in shaping institutional and media attitudes towards migrant settlement in Europe (Stehle, 2006). An illustrative example is the use of the word 'ghetto' in Danish government legislation from 2010 to 2021. As well as officially defining 'ghetto' as any deprived social housing area with a population of more than 50% non-western migrants and their descendants, this legislation introduced countermeasures such as bans on further settlement by non-whites, additional penalties for residents convicted of crimes, and wholesale eviction, demolition and redevelopment if conditions did not improve (Fallov & Birk, 2022; Risager, 2023).

The important question, therefore, is not whether a neighbourhood meets the arbitrary criteria for being defined as a 'ghetto' or 'no-go' area, but the reasons why a place is labelled as such and the consequences of this process for its residents. Loïc Wacquant (2007), Tom Slater (2021) and others have developed the concept of 'territorial stigma' to draw critical attention to the intersection of spatial denigration and material deprivation in the reproduction of urban inequalities, and to the symbolic power that underpins tropes such as 'ghettos', 'parallel lives' and 'self-segregation' that are evoked in public debates about ethnic segregation. Residents typically resist the external stigmatization of their area by, for example, reappropriating the term 'ghetto' as a signifier of neighbourhood pride and collective solidarity (Slater & Anderson, 2012). However, as the Danish 'ghetto laws' demonstrate, territorial stigma has an undeniably negative impact on the economic, social and psychological wellbeing of the targeted populations and ultimately on societal expectations of their 'integration' (Slater, 2021: 169–172).

Armed with evidence-based knowledge and critical insights, academics have intervened in public debates to debunk misinformation and shoddy analysis and to shine the light back on the underlying inequalities at the heart of the contemporary city. Indeed, segregation scholars have been among the most politically and publicly engaged of all researchers working on migration and cities, even when this has meant using the much-maligned category of ethnicity. Nevertheless, academics are not immune to the ideological imprint of a given era, and as Deborah Phillips notes, certain value-laden representations of ethnic segregation have sometimes 'crept into scholarly analyses' (2007: 1144), particularly in the 'neighbourhood effects' literature, which analyses how people's life chances are affected by where they live, over and above the structural and political factors responsible for urban inequality (Slater, 2021: 109–136).

4.3.5 Contested Neighbourhood Identities, Residential Mobility and Transnationalism

Neighbourhoods associated with particular groups—such as the 'Banglatowns' of London and Rome—are far more heterogeneous than is often acknowledged. Three general points are worth highlighting. First, members of the same ethnic or national group differ in terms of age, class, gender, time of arrival, legal status, diasporic affiliations, social and political attitudes, links to other neighbourhoods, cities and countries, and so on. These differences intersect to shape a variety of residential dynamics. For example, the composition of the Bangladeshi population in Tower Hamlets has changed in recent years as higher-income households have relocated to better housing in outer London, while significant numbers of Italian Bangladeshis from Rome and northern Italy have moved to the borough as part of onward migration projects, albeit at the cost of deteriorating housing conditions and, in some cases, tensions with local British Bangladeshis (Della Puppa & King, 2019; Morad et al., 2021).

Second, while attention may focus on a numerically significant ethnic minority, urban areas in Europe have long been characterized by the co-presence of different groups. Drawing on the idea of superdiversity, researchers have explored people's neighbourhood attachments (Pemberton & Phillimore, 2018) and housing strategies (Wessendorf, 2018) in localities where no single group predominates, underscoring, for example, the crucial role of cultural capital and legal status over ethnicity or national origin in influencing the residential trajectories of new arrivals (ibid).

Third, it is important to remember that neighbourhoods are not fixed administrative units like census tracts. Rather, they are protean places where migrants and non-migrants, residents and city users negotiate different boundaries, meanings and attachments. In addition, neighbourhoods experience different levels of population turnover as migrant and non-migrant residents move in and out of areas. This complicates discussions of 'clustering' or 'social mixing'. For example, many central districts of major European cities, including Spitalfields and Torpignattara, which have historically been associated with (and indeed stigmatized for) higher than average concentrations of immigrants, have in recent decades become popular with non-migrant professionals, students and families attracted by their multicultural ambience and affordable housing. The ensuing process of gentrification in such cases can disrupt people's affective ties to place, limit housing options for low-income residents and newcomers, reduce opportunities for the social reproduction of certain communities, and ultimately displace vulnerable households and exacerbate micro-segregation between different groups (Keskiner & Waldring, 2023).

The concern to address the above complexities has driven qualitative research on migrant settlement in cities in recent decades. Particular attention has been paid to residential mobility and transnationalism, both of which have been largely overlooked in segregation studies. In the case of the former, researchers have considered how, when, why and whether migrant households change their place of residence. Like all urban dwellers, migrants' residential mobility may involve short-distance moves within the same area, relocation to another neighbourhood, or longer-distance moves outside the city. Residential mobility is shaped by preferences, opportunities and constraints (desire for distance or proximity to in-group members, access to financial resources, increase in household size, etc.). It may signal an improvement in housing and environmental conditions, or the opposite. It can mean a conscious decision to move into, within or away from an area with a particular demographic profile. Residential *im*mobility, on the other hand, refers to households that remain in the same accommodation for long periods of time, which may indicate that they are stuck due to a lack of resources, but may also imply an enduring attachment to a particular home or location (for a theoretical overview of residential mobility and immobility, see Coulter et al., 2016).

In the UK context, pioneering in-depth interviews with British Asians in Bradford and Leeds, conducted by Deborah Phillips and her colleagues, identified class, gender and generational differences as well as aspirations for housing security, independence and better public services, as key factors in shaping decisions about where to live (Phillips et al., 2007). While many Pakistani and Indian households had left cramped inner-city districts for spacious homes in more ethnically mixed suburbs on the outskirts of both cities, others remained in their old neighbourhoods or relocated to better housing nearby in order to maintain community ties and to avoid the risk of social isolation and harassment in outer areas (for studies of other European cities, see Bolt & Kempen, 2010; Torpan et al., 2022). Residential mobility is also a theme in the emergent literature on highly skilled 'middling' migrants in European cities. Research has considered the interplay of temporal, structural and relational processes that influence and constrain residential trajectories and place attachments in expensive cities such as Helsinki and London (Eskelä, 2018; Maslova & King, 2020; Ryan, 2018). Although professional opportunities, social networks and commitment to long-term projects embed individuals to housing pathways within the same city or neighbourhood, 'processes of dis-embedding' such as relationship breakdowns or exogenous shocks like Brexit can disrupt settlement plans and ultimately lead to onward or return migration (Maslova & King, 2020: 8).

The second strand of research, transnationalism, provides an important lens for thinking about connections and identifications with multiple places that go beyond incorporation into a single, final destination (Gielis, 2009). On the one hand, scholars have explored how migrant settlement in a given city is sustained through the pooling of resources and the exchange of care and emotional support with members of family and social networks in countries of origin and other migrant destinations (Baldassar & Merla, 2014). On the other hand, attention has focused on transnational households that emerge across different places of residence. While recent decades have seen a general increase in multilocal family life, particularly in the West, due to increased occupational mobility and relationship dissolution (Schier, 2016), the concomitant formation of transnational living arrangements among migrants is also mediated by discriminatory migration regimes and levels of socio-economic, legal, gender and racial privilege (Lundström, 2014).

A body of work has examined the transnational strategies of hypermobile elites, from the super-rich who purchase residential property in Europe's largest cities, such as London and Paris, as secure investments and luxury pieds-à-terre that intensify housing and spatial inequalities in these cities (DeVerteuil & Manley, 2017; Fernandez et al., 2016), to northern European middle-class lifestyle migrants who acquire second homes on the coast, in the countryside and, increasingly, in the major cities of Mediterranean countries (Domínguez-Mujica et al., 2021; Gustafson, 2009; Montezuma & McGarrigle, 2019). Meanwhile, many settled labour migrants in western and southern Europe use remittances to buy or build homes in their countries of origin, which can lead them to accept frugal lifestyles and poor housing in their cities of residence, but also cause friction when authorities restrict welfare entitlements due to their ownership of foreign assets (van der Horst, 2010). Moreover, households may be scattered across multiple localities due to limited mobility options, with some more accessible cities becoming nodes in multi-stage strategies to secure economic and legal stability (McGarrigle & Ascensão, 2018).

4.3.6 The Intimate Sphere of Settlement: Migrant Homemaking in the City

In the last decade, research on segregation, residential mobility and transnationalism has been accompanied by a growing interest in migrant 'homemaking'. Rather than viewing home as a fixed place of shelter, scholars consider homemaking to be an open-ended, relational process that spans household, community, city and nation.

This approach provides valuable insights into the living conditions and life projects of migrants, as well as their creation and maintenance (or loss) of ties to people and places (Boccagni, 2023; Fathi & Ní Laoire, 2024).

By exploring the subjective dimensions of migrant settlement, scholarly work on homemaking fills a crucial gap in a literature long dominated by 'disembedded', statistical data-driven approaches (Picker, 2017: 5–7). Although much of this research has been conducted in urban settings, only some of it is directly useful for thinking about the relationship between migration and cities. It is possible to identify two general foci: first, work that interrogates the distinction between the intimate space of the home and the 'outside' world of the urban environment, and second, the process of homemaking in the city itself.

In the first case, micro-level research has drawn attention to the often ambivalent and porous relationship between the domestic sphere and the outside world. While the comforts of home can offer a refuge from the hostility experienced on the city's streets (Back et al., 2018: 144), the confines of home can also be associated with overcrowding, lack of privacy and intergenerational conflict. Some of the young Asian women interviewed in Phillips et al.'s (2007) study of Bradford and Leeds expressed a desire to move to the edge of, if not out of, central neighbourhoods in order to reduce the hospitality demands placed on them to maintain 'open houses.' Moreover, it was the home, rather than the city, that was sometimes a place of insecurity. Indeed, poor multigenerational Asian households in inner-city neighbourhoods were among the hardest hit groups in the UK during the Covid-19 pandemic (Boseley, 2021).

The boundary between domestic and non-domestic space is drawn in profoundly different ways. As Paolo Boccagni and Andrea Mubi Brighenti argue, migrant homemaking can be fruitfully understood in terms of shifting 'thresholds of domesticity', where the passage between private, communal and public space is 'fragmented, discontinuous and even reversible' and characterized by 'tensions and frictions' (Boccagni & Brighenti, 2017: 4–5). In his ethnographic study of 'Hotel House', a high-rise former holiday complex on the Italian Adriatic coast inhabited by more than forty national groups, Adriano Cancellieri (2017) considers collective homemaking practices in the condominium, including the creation of an Islamic prayer room and small shops on the ground floor. While these practices contributed to the creation of 'homely relations' and communal bonds, they also revealed a 'dark side of home' by imposing greater levels of social control, particularly over the activities of women and young people (ibid: 58).

A second strand of research has focused specifically on what it means to build a sense of home in relation to the city. Anabelle Wilkins's (2019) ethnography of Vietnamese migrants in London explores how different feelings of belonging in the British capital are influenced by the time and circumstances of people's arrival, as well as changes in immigration policy, housing markets and employment opportunities. Mastoureh Fathi and Caitríona Ní Laoire (2023) propose the concept of 'urban home' to examine young male migrants' connections and disconnections with the city of Cork, Ireland, a relatively new destination compared to other cities in northern and western Europe. The authors identify four dimensions that shape lived experiences and personal histories of the city: instrumental sites of repeated practices and encounters; places associated with memories of pre-migration life; spaces of care; and the notion of future home. 'Imagined future home', they argue, 'may (not) be about Cork; yet, Cork provides the settings and spaces in which future home is lived and felt in the present' (ibid: 826).

Analyses of homemaking processes, at both the micro-spatial and urban scale, have challenged the binary between sedentariness and mobility that has underpinned research on migrant settlement in cities. In addition, they have highlighted the often-neglected temporal aspects—the connections between past, present and future—that mark migrants' emplacement in and between urban localities.

4.4 Migrant Urbanisms: Using and Remaking the City

The chapter now shifts its focus from urban residential processes to how migrants use and transform the built environment around them. A useful starting point is the general organizing idea of 'migrant urbanisms.' First proposed by the sociologist Suzanne Hall (2015) in her ethnographic study of a London high street, this term refers to the various practices and interactions through which migrants remake urban space, not just to meet immediate needs but to actively redefine what it means to belong in a city. According to Hall, migrant urbanisms unfold within unequal and racialized contexts and as such are also 'about everyday resistance and extend to how resolutions are forged and how disputes are maintained between diverse groups and between those with more power and those with less' (Hall, 2015: 865). Thus, migrant urbanisms not only provide insights into how migrants 'fit' themselves into a city, but through their entanglement with the processes of urbanization, they also point to more general questions about who the city is for, what the city should look like, how it is changed and by whom.

Engaging critically with such questions requires us to grapple simultaneously with two methodological dilemmas that lie at the core of migration studies. First, how are we able to define our object of analysis? How can we tell, for example, if someone in public space is a 'migrant'? Through embedded knowledge or through observable markers such as physical appearance, clothing or language? The latter are, of course, unreliable and ultimately open the door to baseless speculation and racial profiling. For example, at the height of a moral panic about urban insecurity in Italy in the early 2000s, mainstream journalists and politicians typically assumed that people of colour hanging out in the central public spaces of major cities were '*clandestini*'—undocumented immigrants (Dines, 2012: 190–192). But even more measured representations of urban diversity rely on assumptions about a decipherable public, as the map of London discussed at the beginning of this chapter illustrates.

Second, should migrants be studied separately or as part of the general public? Focusing on newcomers or those with irregular status might deepen our understanding of the prohibitive conditions under which they experience urban space, but it could risk amplifying this particular dimension of their everyday lives to the detriment of others. It could also overlook the possibility that non-migrants may face similar mechanisms of control or displacement. As Bridget Anderson has noted, 'while a citizen may have a right to be present on the territory that does not give them the right to be in any public space. Citizens who are homeless or who beg can be prohibited access to certain spaces or moved on' (2019: 10).

With these caveats in mind, I consider here two key domains of migrant urbanism: open public space and built social infrastructure. This distinction is made for practical purposes and reflects different areas of interest in the literature. In reality, open and built spaces are often closely linked, both in terms of how publicness is theorized at the urban scale, and in terms of the spatially interconnected nature of everyday life in cities.

4.4.1 Open Public Space

Like all people, migrants use and adapt open public space in a variety of ways: to navigate the city, to meet acquaintances and strangers, to be alone, for leisure activities, for economic transactions, for protest, and so on. Streets, squares, parks and other outdoor spaces are a fundamental part of the social and symbolic fabric of cities. As sites of difference, display, proximity and interaction, they are also imbued with political and ideological significance and are routinely implicated in disputes over access, surveillance, rights, security and propriety. Such issues are in turn shaped by socio-demographic factors including gender, race, class and age; shifting diagnoses of what is at stake; the location, type and status of the space itself; and the 'public space cultures' (intensity of use, local norms, etc.) that vary from city to city.

Open spaces offer compelling settings for investigating different dimensions of migrants' relationships with cities, whether we approach public space from a 'topographical' perspective as located in particular places, or from a 'procedural' perspective as a dimension that can potentially be brought into being anywhere (Iveson, 2007: 1–20). From the bustling central square to the pavement outside the school gates, different open spaces illuminate the opportunities and networks that contribute to building urban livelihoods, as well as the barriers and conflicts that need to be negotiated. They tend to generate granular and situated rather than generalizable information. Some migrants may use public space extensively, perhaps because they live in inadequate accommodation and spend most of their time outdoors, or more sparingly because of experiences of discrimination. Certain practices in some spaces may come under public scrutiny and therefore be susceptible to regulation, while in others they may remain inconspicuous.

The literature on migration and public space in European cities can be divided into three main areas of inquiry: first, migrants' multiple uses and reconfigurations of public space; second, the legislative and public-discursive contexts in which this occurs; and third, the links between migrants in public space and broader processes of urban development. Although these three areas often overlap, especially in in-depth case studies, it is helpful to consider them separately.

The first area of inquiry characterizes the bulk of the literature. Researchers have explored the functional, social, economic and symbolic affordances of different types of open space for migrant and minority groups, from parks (Rishbeth et al., 2019) to local shopping streets (Hall, 2015; Zukin et al., 2016). Others have looked at combinations of sites to examine specific dynamics of migrant incorporation: the acquisition of practical knowledge about a city and its resources (Buhr, 2018); the construction and maintenance of spatial boundaries (Noussia & Lyons, 2009); the formation of affective attachments to place (Rishbeth & Powell, 2013); and the establishment of translocal connections through, for instance, makeshift postal services (Dines, 2012). Numerous studies have drawn attention to the intersections of gender, race, religion and age in structuring migrants' experiences of public space (Díaz-Cortés et al., 2008; Ehrkamp, 2008; Schmoll, 2019), not only to highlight the complexity of urban diversity but also to affirm the inherently contested nature of public space where, for example, divergent notions of socialization and morality come into contact. (On the relationship between public space and migrant politics, see Sect. 6.1).

The second strand of research focuses its attention on contextual factors—the policies, technologies, modes of representation, etc.—that shape the relationship between migration and public space. Researchers have explored, among other things, the impact of migration policies on the creation of internal border controls in cities (Fauser, 2019), the racialization of migrants and the policing of their bodies in public (Amer & Leung, 2023), and institutional attempts to incorporate celebrations of diversity into the design of urban parks (Stanfield & van Riemsdijk, 2019). Others have analyzed news coverage of public spaces associated with migrants and minority groups, particularly sites prone to stigmatization such as transport hubs, and how media frames influence local policy decisions (Cancellieri & Ostanel, 2015). Scholars simultaneously consider how migrants negotiate and resist the various interventions and representations targeted at them, for example through invisibilization strategies such as dressing and behaving like tourists (Amer & Leung, 2023).

The third body of work considers the interconnections between public space and broader processes of urban change. The relationships that migrants form with public space sometimes have implications far beyond the actual places themselves. Migrants can contribute to economic growth, for example by revitalizing street life (Çağlar & Glick Schiller, 2018), but they can also be perceived by local authorities as an obstacle to commercial redevelopment (Obradovic-Wochnik, 2018). My own ethnographic research on Piazza Garibaldi in front of Naples central railway station—an important social, economic, religious and political space for various migrant groups since the 1980s—explored how the vast square became entangled in the politics of urban regeneration (Dines, 2002, 2012). Regeneration in Naples was more than a development strategy aimed at attracting foreign investment and tourism. It was a symbolic and ideological project through which the city's centre-left administration sought to rid Naples of its international reputation for urban decay, poverty and organized crime, and to redefine the relationship between its inhabitants and the public realm. Like many station areas in Europe, Piazza Garibaldi had long been seen as chaotic and run-down, but rarely as newsworthy in and of itself. With the introduction of a public narrative of urban renewal in the 1990s, the pre-existing presence of migrants in the square was problematized as a source of disorder and insecurity, and subject to increased control. Although migrants had a marginal voice in the public debate on regeneration and were forced to deal with the material consequences of its symbolic regime, they nevertheless played a fundamental role in the creation of a new Piazza Garibaldi and were instrumental in the revival of the area's wholesale economy, which served to reposition Naples at the centre of trans-Mediterranean networks of petty trade.

The links between migration and open public space persist over time and become embedded in meanings of place, but they also change according to the shifting tectonics of international mobility and urbanization. In the case of Piazza Garibaldi, changes such as the arrival of asylum seekers and refugees in Naples since the 2010s and the end of the previously aggressive regeneration agenda have set the stage for new research, such as the multilingual interactions and ambivalent solidarities between local and migrant street vendors (Dawes, 2020). In sum, open public space remains an important portal to the deeper dynamics at play in a city, whether for tracing processes of migrant emplacement or for understanding the impact of international migration on urban restructuring.

4.4.2 Social Infrastructures and Religious Spaces

Like everyone, the daily lives of migrants in cities are also mediated through a range of built spaces characterized by different levels of access and ownership, from cafes (Ehrkamp, 2005) and sports clubs (Janssens & Verweel, 2014) to public transport (Wilson, 2011) and libraries (Robinson, 2020). Such places can usefully be understood as 'social infrastructure' (Latham & Layton, 2019). This term essentially refers to what enables a city to function socially; in other words, the spatial arrangements and associated situations and events that facilitate activities and encounters that potentially lead to more extensive use of the city, the building of social networks, cooperation and trust, and can provide important relief in times of stress. Alan Latham and Jack Layton (2019) identify five overlapping types of social infrastructure: public institutions, commercial enterprises, recreational venues, religious spaces, and transit-related infrastructure. Thinking infrastructurally draws attention not only to the affordances of material settings, but also to the practices that produce a sense of publicness and the relationships established between people and places. For example, the regular users of a local library transform the building into more than just a place of study by forming friendships and communities of care and appropriating spaces on its premises where these can be sustained. The ways in which such spaces are produced, adapted, shared and contested depend on various factors: the flexibility of rules, the openness to difference, the quality of the environment, the level of maintenance, the distribution of resources, and so on. It also means taking into account what has been called 'infrastructural violence': the unequal provision and withdrawal of social infrastructure in cities, which has a disproportionate impact on the lives of low-income and disadvantaged users (ibid: 2).

As migrants settle in a city, they inevitably come into contact with its pre-existing social infrastructures, including those specifically designed to facilitate their 'integration' (Kohlenberger, 2024). During this process, they learn to differentiate between different facilities on the basis of their benefits and drawbacks. While some infrastructures may play a central role in everyday life, others may be avoided because of their social exclusivity or simply ignored because they are of no interest. In essence, this reflects a common experience of all city dwellers. However, migrants also produce their own infrastructures that respond to specific needs and desires. These include officially recognized establishments such as restaurants, barber shops, community centres and places of worship, as well as more informal, improvised or temporary spaces such as prayer rooms, street stalls and squatted buildings. In all cases, the infrastructural character is enacted through the practices and social relations created in situ. In addition to being places for social gatherings and specific activities, these migrant social infrastructures contribute to the formation of group identities, create conditions for self-organization and representation, act as sources of information and assistance for newcomers, and support translocal mobilities through their connection with people and places elsewhere (Jung & Buhr, 2022; Meeus, 2017; Wessendorf & Gembus, 2024). Infrastructural practices also involve acts of 'vernacular creativity' (Gilbert et al., 2019) that transform the built environment, for example through the redesign of buildings and the production of new linguistic landscapes (Blommaert, 2014). These visible changes often contribute to placemaking processes and can be selectively integrated into branding strategies outlined in the previous chapter (Sect. 3.6), but in areas of cities with long and diverse histories of migration, they may be so commonplace as to appear unremarkable (Hentschel & Blokland, 2016).

While infrastructures are often associated with particular national, ethnic or religious groups, they do not have the same meaning for all group members. For example, the numerous Turkish teahouses studied by Patricia Ehrkamp in the working-class district of Marxloh in Duisburg catered to different people depending on their regional origin, sports club affiliation and political orientation, but all were ultimately geared to the needs of the male immigrant population (Ehrkamp, 2005: 344–345). A neighbourhood may host multiple infrastructures that reflect class, gender, political and religious fault lines within a single 'community' (Baumann, 1996; Phillips et al., 2007). Some infrastructures reproduce rivalries in sending societies, generate new power dynamics in diasporic communities, or find themselves at the centre of tensions with neighbouring residents, especially when they are perceived as disrupting the social order or violating local meanings of place (Pastore & Ponzo, 2016).

A growing body of research has explored the relationship between religion and migrant incorporation in cities (Beaumont & Baker, 2011; Garbin & Strhan, 2017; Garnett & Harris, 2013; Oosterbaan, 2014). Over the last three decades, international migration has played a crucial role in reviving and diversifying religion in urban life, particularly in northern and western Europe, where it had previously appeared to be in decline, which has fuelled debates about the emergence of 'post-secular cities' (ibid). Migrant worshippers have joined, and often reanimated, existing congregations in cities, but have also established their own religious institutions. New churches, mosques and other places of worship vary widely in visibility, size and permanence, ranging from converted garages in apartment blocks to purpose-built structures that incorporate non-local architectural elements (Chiodelli, 2015; Eade, 2017; Gilbert et al., 2019). They exist alongside a network of faith-based services, such as halal butchers and Sikh community kitchens, and a busy calendar of public ceremonies and festivals. In addition to meeting spiritual needs, these assorted religious spaces represent hubs of communal life and important sources of charitable welfare for migrants and non-migrants alike (Vine, 2024).

Today, religious diversity is both a defining feature of major global metropolises (Garbin & Strhan, 2017) and increasingly evident in smaller cities and peripheral areas of Europe (Martikainen et al., 2019). In recent years, multi-faith institutions and services have spread from traditional 'arrival' neighbourhoods to suburban areas, largely driven by a combination of residential mobility, less stringent planning regulations and reduced public scrutiny (Dwyer et al., 2013; Gilbert et al., 2019).

These developments are not without controversy. While some of this is the inevitable result of competition between different interest groups for scarce urban resources such as property and land (Eade, 2017), a major recurring flashpoint has been the hostility of members of majority populations to religious diversity itself. The greatest hostility has been directed at Islam, articulated in particular through organized opposition to new mosques (Cesari, 2005; Göle, 2016). Such actions have involved local residents and city councillors as well as national politicians and social media influencers, and have typically mobilized around a mix of grievances such as the location, size and style of the building, its perceived impact on traffic and house prices, and preconceptions of Islam as a threat to cultural values and security (Astor, 2016; Gale, 2004). Some of the most intense protests have occurred in cities with a relatively recent history of international migration, especially in southern and eastern Europe, and/or where there are established groups and organizations that politically mobilize around anti-immigration and anti-Islam platforms (Astor, 2016; Saint-Blancat & Schmidt di Friedberg, 2005), regardless of whether Muslims have longstanding and uncontroversial ties to the same cities and their respective countries (Narkowicz & Pędziwiatr, 2017).

What is at stake is often more than a refusal to accept the public presence of Europe's largest religious minority: criticism of Islam also serves as a foil for racism and xenophobia and reflects wider tensions in contemporary urban societies, such as conflicts between different levels of governance and power struggles between secularism and religious pluralism (Berking et al., 2018). Anti-mosque campaigns have been consistently successful, thanks to the capitulation of local authorities, but also to Muslim groups withdrawing or significantly modifying their proposals to avoid further confrontation (Saint-Blancat & Schmidt di Friedberg, 2005), although in cities with consolidated and well-organized Muslim communities, particularly in the UK, France and Belgium, public opposition has waned considerably. In the meantime, overt anti-Islamic sentiments have been absorbed into ethno-nationalist populist politics, sometimes erupting in urban violence, as in the spate of racist pogroms in towns and cities across England and Northern Ireland in the summer of 2024. However, if nativist resentment against migrant religious infrastructures points to prejudices embedded within European societies such as Islamophobia and the conflation of official forms of Christianity with whiteness and Western civilization (Lauwers, 2023; Najib & Hopkins, 2020), day-to-day coexistence with religious diversity in metropolitan and suburban contexts is invariably a more mundane, ambivalent and inconclusive process than the binary perspectives that emerge from public controversies (Baker, 2019). It is to the question of everyday social interaction and conflict that the final section of this chapter now turns.

4.5 Living with Difference: Conviviality, Racism and Conflict

Migration not only transforms the spatial and built environments of cities, but also reorganizes the social relations that take place within them. In the early 1990s, the Jamaican-born British cultural theorist Stuart Hall claimed in an often-quoted comment that the 'capacity to live with difference' would be 'the coming question of the twenty-first century' (Hall, 1993: 361). The same question had long preoccupied Western sociological thinking about the urban condition, from Georg Simmel's (1950 [1908]) reflections on encounters with strangers in urban life, through Chicago School ethnographies of market-mediated inter-ethnic relations such as Paul Cressey's *The Taxi-Dance Hall* (1932), to later studies of cross-cultural communication, particularly among young people (Hewitt, 1986). For the European social sciences at least, Hall's prediction seems to have been correct: the twenty-first century has seen a surge in scholarly interest in people's encounters with difference (Colombo & Semi, 2007; Darling & Wilson, 2016; Gilroy, 2004; Rogaly, 2020; Wise & Velayutham, 2009). This interest is partly driven by the intensification and diversification of migration brought about by globalization, but it is also a reaction to, and sometimes openly at odds with, conventional ethnic group-based migration research, perfunctory references to cultural diversity, and policy paradigms such as social integration and community cohesion that single out neighbourhoods and groups of residents as disconnected from mainstream society (Karner & Parker, 2011).

Most recent theoretical and empirical work on urban multiculture has critically distanced itself from prescriptive and affirmative understandings of social interactions (Valentine, 2008; Wilson, 2017). Encounters with difference are not an inevitable consequence of culturally diverse cities, nor are they necessarily meaningful or desirable for all concerned. They are the product of social practices, identities and relations situated within wider urban social processes. Hence, the fact that migrant and non-migrant parents may intermingle outside the school gates is not simply the result of their children being in the same class, but also the result of local socio-economic, housing and institutional conditions that determine how different families live within the same catchment area. People may be more or less willing or able to mix and bond with diverse others. Newcomers to an urban area sometimes actively seek to build bridging social capital across ethnicity and class to secure a foothold in local life; wealthy transnational elites reserve the prerogative to avoid interaction at all costs; while self-identified members of the 'native' white working class may, by virtue of their daily routines, enjoy more meaningful interpersonal relations with people with migration backgrounds than members of the ostensibly cosmopolitan-minded liberal middle class who inhabit largely homophilous social worlds (Rogaly, 2020; Vincent et al., 2018). Migration and ethnicity may appear to have little bearing on the formation of social ties, for example in cases where length of residence, participation in local activities and adherence to dominant norms and values are considered more consequential, and where non-compliance is associated with other types of 'outsiders' such as gentrifiers, students and tourists (Wimmer, 2004). Everyday sociality in multicultural cities is thus highly ambiguous. It crosses but also reconstitutes social boundaries, is both inclusive and exclusive, and plays out against a backdrop of structural and racial inequalities.

A number of scholars have examined the organization of encounters at institutional and grassroots levels, assessing the transformative potential and unintended consequences of various initiatives (Christiansen et al., 2017) and addressing the power asymmetries through which difference is defined, enacted and valued (Stansfeld, 2024). The bulk of research, however, has focused on *un*organized interactions with difference and has led to the development of a number of interchangeable concepts such as 'everyday multiculturalism' (Colombo & Semi, 2007; Wise & Velayutham, 2009), 'throwntogetherness' (Gawlewicz & Yiftachel, 2022), 'non-elite cosmopolitanism' (Rogaly, 2020), 'commonplace diversity' (Wessendorf, 2014) and 'conviviality' (Gilroy, 2004; Wise & Noble, 2016). In particular, the latter concept has been widely used over the past two decades to reflect on 'the process of cohabitation and interaction which have made multiculture an ordinary feature of [urban] life' (Gilroy, 2004: xi). The analytical power of conviviality, and the core of Paul Gilroy's seminal formulation, is that it does more than simply describe mundane practices of interaction: it points to how these unfold in the context of—and in spite of—racism, anti-immigrant sentiment and social insecurity (De Noronha, 2022; Valluvan, 2016; Wise & Noble, 2016). Thus, when used critically, conviviality 'should not be a by-word for saccharine diversity fantasies but the difficult work of understanding how people live with each other peacefully in the midst of the political ruins of racism' (Back et al., 2018: 134). In the case of Les Back and Shamser Sinha's co-research with young migrant residents in London, this 'difficult work' involved identifying common tools—namely 'fostering attentiveness and curiosity', 'care for the city and a capacity to put yourself in another's place', 'worldliness and making connections beyond local confines', 'developing an aversion to the pleasures of hating' and 'creating local networks and building a home'—that enable people to produce spaces of coexistence in conditions of adversity (ibid: 135–145). Back and Sinha provide a range of examples including the story of an Afghan man living in an east London council flat who, while waiting for his asylum claim to be processed, cultivated ties with his initially hostile elderly neighbours by building them a bench from discarded wood found on his street. Ten years earlier, he had used these same carpentry skills to build a boat to cross the Mediterranean Sea.

Alongside a lively theoretical and methodological debate (see also Lapina, 2016; Padilla et al., 2015), there is an extensive body of work that examines conviviality in concrete contexts. This has explored interactions both between people with migration backgrounds and members of the white majority society (Koulaxi, 2022; Tyler, 2017) and between members of different migrant, refugee and minority groups (Rosbrook-Thompson, 2018), also across the intersections of age, class and gender (Nayak, 2017), and in some cases considering how encounters lead to acts of racism (ibid) or to enduring forms of friendship and collaboration (Krifors, 2022). Ethnographic research has been conducted in a variety of urban settings, from inner-city social housing (Back et al., 2018) and middle-class suburbs (Tyler, 2017) to small provincial cities (Rogaly, 2020), and has looked at a range of micro-publics, urban spaces and infrastructures, from the mainstays of social research such as street markets and cafes (Dines, 2009; Samanani, 2023) to less usual sites such as a bowling alley facing demolition (Jackson, 2019), a football club in a superdiverse neighbourhood (Rosbrook-Thompson, 2018) and an immigration removal centre (De Noronha, 2022). Although most work published in English focuses on the UK, the concept of conviviality in relation to migration has been explored in different urban settings across Europe—including a multicultural district in central Athens (Koulaxi, 2022), a street market in Sofia (Venkov, 2019) and small towns in rural Sweden (Krifors, 2022)—while the use of the term as a critical concept for dealing with everyday encounters with difference has been developed in other, particularly Latin-based, languages (e.g. Marabello & Riccio, 2020).

Conviviality, many scholars insist, takes place in the face of conflict and discomfort. It is not, as the English definition suggests, about 'festive togetherness', nor is it reducible to normative notions of good community relations. Conversely, an absence of social interaction does not necessarily imply a lack of integration. By its very nature, the city allows for non-involvement and indifference. Early urban sociologists such as Georg Simmel and Louis Wirth famously argued that people cope with the 'stimulus overload' of metropolitan life by disengaging and ignoring others in public (Lofland, 1998). Erving Goffman later identified rituals of 'civil inattention' that underpinned an 'interaction order' through which city dwellers and users largely self-regulated their face-to-face encounters with strangers, allowing for minimal recognition while sparing people the need to either avoid or meaningfully engage with others (Goffman, 1963). The elementary point here, often lost on policymakers and migration researchers alike, is that newcomers to cities *also* seek to control their encounters with different publics. Many southern Italian families who migrated to northern Italian cities in the 1950s and 1960s aspired to eventually move into apartments in new peripheral public housing blocks where the negative dimensions of social isolation were offset by the prospect of greater privacy, modern comforts and 'liberation' from the intrusive environment of tenement courtyards and hometown squares (Foot, 2001: 53–59). Similarly, first and second generation South Asian women in Bradford, England, have moved away from inner-city neighbourhoods to strike a better balance between support networks on the one hand and personal independence and freedom on the other (Phillips et al., 2007). Such examples call for a more nuanced understanding of migration and urban sociality, one that takes into account not only the enduring and fleeting instances of conviviality, but also the ritualized forms of public behaviour, voluntary acts of avoidance and experiences of anonymity that are fundamental to the everyday social functioning of cities and that make urban life both attractive and unsettling.

Overlapping with these dispositions are external factors that constrain and mediate interactions or create tensions and frictions between migrants and others. Many of these factors are not directly related to migration, such as the privatization of urban space, the expansion of surveillance systems, or the rapid development of smartphone technology that changes the way people interact in and with the built environment (Brighenti & Pavoni, 2022). Some issues, such as personal safety and access, are cross-cutting in terms of gender, age and physical ability, but can be exacerbated by a person's migration status, while others, such as language barriers, are more closely linked to migration, though not necessarily coterminous with it. Tensions arising from competition for shared spaces and common resources can delimit meaningful interactions between migrants, including members of the same national group, as exemplified by standard Bangla-speaking Bangladeshi Italians in the East End of London who have carved out their own enclaves of multicultural sociability within the broader context of the local Sylheti-speaking Bangladeshi community (Della Puppa & King, 2019). Moreover, as noted above, regional and political differences within migrant groups can lead to the reproduction of divisions and rivalries on arrival, while particular uses of public space by certain group members can work to exclude (weaker) others, neither of which augurs well for the cultivation of conviviality.

At the same time, it is important to reiterate that any serious understanding of what it means to live with difference in contemporary European cities cannot underestimate the 'enduring significance of racism' (Gilroy, 2004: 135). Operating at multiple levels—structural, political, institutional, societal, blatant, subtle and so on—racism deeply shapes the social experience of migration-driven diversity. The racial hierarchies embedded in international migration management are reproduced through unequal access to housing, work, education and services, which in turn affects the possibility, need and desire for social interactions. The combination of racial profiling and stop and search practices associated with immigration enforcement checks throughout Europe (De Maillard et al., 2024) transforms everyday sites of sociality, such as train stations and busy street corners, into potential danger zones and heightens feelings of fear, anger and distrust among those who see themselves as targets (Jones et al., 2017). Racism also plays out at a more mundane but no less pernicious level through the everyday encounter and negotiation of racialized boundaries of belonging and claims to localness (Gruner, 2010; Nayak, 2017).

People's mental maps of urban diversity draw not only on knowledge of group distribution or migrant amenities, but also on place-specific experiences and perceptions of discrimination. British scholars, for example, have examined how white-dominated residential areas on the edges of multicultural cities are imagined and performed by some locals as 'last bastions of whiteness' where encounters with difference are marked by animosity and harassment (Hardy, 2017; Nayak, 2010: 2377). Migrant neighbourhoods themselves, including those branded as tourist attractions, often conceal histories of racist violence perpetrated by people claiming to defend local identities and incited by supremacist ideologies. The London and Rome districts discussed earlier in this chapter were both the scenes of racist murders: the Bangladeshi textile worker Altab Ali, who was stabbed by three teenagers near Brick Lane in 1978, at a time when the area was a target of the far-right National Front (Alexander, 2011) (Sect. 6.6), and the Pakistani Muhammad Shahzad Kahn, who was beaten to death on a street in Torpignattara in 2014 by a local white youth irritated by the man's erratic behaviour (Santoro, 2015).

Acts of racial violence can exacerbate socio-spatial divisions and leave scars in collective memory. They can also lead to collective responses, including incidents of major urban disorder. The latter—commonly referred to as 'riots'—have often been triggered by state violence in the form of police harassment and killings, and are invariably underpinned by accumulated experiences of racial disadvantage and institutional mistreatment. Over the past forty years, such events have tended to involve young people with migration backgrounds, sometimes flanked by 'native' white youth, rather than foreign-born migrants. There is a well-established sociological and criminological literature on urban riots in European cities, particularly in Britain, France and, more recently, Sweden (Body-Gendrot, 2013; Schierup et al., 2014; Waddington et al., 2009) that examines the aetiology and life cycle of events (for a review, see Newburn, 2021), the reasons why they occur in some cities and countries and not others (Lukas, 2009; Mitchell, 2011), as well as public and political responses to them, including how locally-born rioters come to be recast as alien others (Gilroy, 2004: 134). Scholars have also explored the politicalness of riots, for example how they sometimes generate precise demands, become catalysts for alternative narratives about place, and play an important role in the formation of anti-racist and class consciousness across different groups (Schierup et al., 2014).

Experiences of racial violence do not always lead to further violence. They have also activated instances of organized resistance, exemplified by the south Asian youth movements that emerged in British cities in the early 1980s (Ramamurthy, 2013), and created moments of inter-group solidarity, as in the cases of Brick Lane and Torpignattara, in which ethnic and national differences are superseded by other forms of collective identification (Back et al., 2018; Baumann, 1996). However, in cities characterized by rising levels of inequality and structural and institutional violence, where meaningful channels of democratic participation have closed or disappeared, and where migration is publicly scapegoated for social problems and racialized divisions persist, the potential for urban violence around the issue of migration or involving people with migration backgrounds always remains a possibility.

4.6 Conclusion

Through their settlement, migrants transform the spatial and social relations of cities. This chapter began with a juxtaposition of a map celebrating London's diversity, printed in *The Guardian* newspaper, and the Brexit Party leader's cursory disapproval of the same places. Despite their different tones, I suggested that the two examples shared similar characteristics. In particular, they both highlighted the routine manner and apparent ease with which migration is recognized, located and labelled in urban contexts. Any single snapshot or representation on a map—however conceptualized or data-validated—cannot capture the intersection of the multiple temporal and spatial processes and scales that shape any given situation, and, as such, are inevitably partial and invariably political.

This chapter has explored the complex processes that unfold after migrants arrive in cities. It has considered a vast literature that crosses disciplinary boundaries, adopts different methodological approaches and is often actively engaged in public debate. The multiplicity and complexity of the issues at stake is reflected in the fact that this chapter is longer than others. While there is little point in attempting a definitive synthesis, three key themes running through the discussion should be highlighted. First, migrants arrive and settle in cities that are already characterized by different levels of inequality and urban restructuring that affect the population as a whole, and therefore issues such as housing and public space need to be examined more carefully, both in terms of pre-existing conditions and the ways in which these same conditions have been shaped by migration. In other words, this means repeatedly *demigrantizing* the analysis of migrant settlement in order to avoid the trap of approaching it as a separate, discrete object of research (Anderson, 2019). Second, the importance of unpacking the conditions of migrant incorporation extends to thinking about the divergent, context-dependent paths between European cities. The powerful imaginary of area-based ethnic segregation—closely associated with certain northern European cities—is not helpful in making sense of vertical stratification in parts of central Naples where recently arrived migrants and the local middle class live in the same buildings. Third and finally, research on settlement processes needs to critically control its use of catch-all categories and concepts and reflect on the consequences of using them. Unless qualified, 'diversity' can mean everything and nothing. A passing reference to 'segregation' may be laden with unspoken assumptions about the position and place of migrant and minority groups in cities. At the same time, 'ethnic segregation' does not inevitably imply succumbing to the pitfalls of ethnic groupism but can be used as a powerful tool for critique.

References

Alexander, C. (2011). Making Bengali Brick Lane: Claiming and contesting space in East London. *The British Journal of Sociology, 62*(2), 201–220.

Amer, A., & Leung, M. (2023). Not-so-subtle subtleties: Undocumented migrant (in)visibility, (im)mobility and Dutch public spaces as sites of embodied racialization. *Journal of Ethnic and Migration Studies, 49*(15), 3940–3957.

Andersen, H. S. (2017). Selective moving behaviour in ethnic neighbourhoods: White flight, white avoidance, ethnic attraction or ethnic retention? *Housing Studies, 32*(3), 296–318.

Anderson, B. (2019). New directions in migration studies: Towards methodological de-nationalism. *Comparative Migration Studies, 7*(36). https://doi.org/10.1186/s40878-019-0140-8

Arbaci, S. (2019). *Paradoxes of segregation: Housing systems, welfare regimes and ethnic residential change in southern European cities*. Wiley.

Arbogast, L., & Migreurop. (2016). *Migrant detention in the European Union: A thriving business*. Rosa Luxemburg Stiftung.

Astor, A. (2016). Social position and place-protective action in a new immigration context: Understanding anti-mosque campaigns in Catalonia. *International Migration Review, 50*(1), 95–132.

Back, L., & Sinha, S. with Bryan, C., Baraku, V., & Yembi, M. (2018). *Migrant city*. Routledge.

Baldassar, L., & Merla, L. (Eds.). (2014). *Transnational families, migration and the circulation of care: Understanding mobility and absence in family life*. Routledge.

Baker, J. (2019). "Is it a mosque?" The Islamization of space explored through residents' everyday "discursive assemblages." *Identities, 26*(1), 12–32.

Baumann, G. (1996). *Contesting culture: Discourses of identity in multi-ethnic London*. Cambridge University Press.

Beaumont, J., & Baker, C. (Eds.). (2011). *Postsecular cities. Space, theory and practice*. Continuum.

Benedictus, L. (2005, January 21). Every race, colour, nation and religion on earth: Part 1. *The Guardian*. https://www.theguardian.com/uk/2005/jan/21/britishidentity.race

Berking, H., Steets, S., & Schwenk, J. (2018). *Religious pluralism and the city: Inquiries into postsecular urbanism*. Bloomsbury.

Bernt, M., Hamann, U., El-Kayed, N., & Keskinkilic, L. (2022). Internal migration industries: Shaping the housing options for refugees at the local level. *Urban Studies, 59*(11), 2217–2233.

Blommaert, J. (2014). Infrastructures of superdiversity: Conviviality and language in an Antwerp neighborhood. *European Journal of Cultural Studies*, *17*(4), 431–451.

Boccagni, P. (2023). Introduction: Home and migration—Setting the terms of belonging and place-making on the move. In P. Boccagni (Ed.), *Handbook on home and migration* (pp. 1–28). Elgar.

Boccagni, P., & Brighenti, A. M. (2017). Immigrants and homes in the making: Thresholds of domesticity, commonality and publicness. *Journal of Housing and the Built Environment, 32*(1), 1–11.

Body-Gendrot, S. (2013). Urban violence in France and England: Comparing Paris (2005) and London (2011). *Policing & Society, 23*(1), 6–25.

Bolt, G., & van Kempen, R. (2010). Ethnic segregation and residential mobility: Relocations of minority ethnic groups in the Netherlands. *Journal of Ethnic and Migration Studies, 36*(2), 333–354.

Boseley, S. (2021, April 30). UK's south Asian communities worst-hit in Covid second wave, study finds. *The Guardian*. https://www.theguardian.com/world/2021/apr/30/uks-south-asian-communities-worst-hit-in-covid-second-wave-study-finds

Brighenti, A. M., & Pavoni, A. (2022). Goffman back in town: On new relations in public. *Etnografia e Ricerca Qualitativa, 15*(1), 93–114.

Buhr, F. (2018). Using the city: Migrant spatial integration as urban practice. *Journal of Ethnic and Migration Studies, 44*(2), 307–320.

Burgess, E. (1928). Residential segregation in American cities. *The Annals of the American Academy of Political and Social Science, 140*, 105–115.

Çağlar, A., & Glick Schiller, N. (2018). *Migrants and city-making: Dispossession, displacement and urban regeneration*. Duke University Press.

Cancellieri, A. (2017). Towards a progressive home-making: The ambivalence of migrants' experience in a multicultural condominium. *Journal of Housing and the Built Environment, 32*(1), 49–61.

Cancellieri, A., & Ostanel, E. (2015). The struggle for public space: The hypervisibility of migrants in the Italian urban landscape. *City, 19*(4), 499–509.

Castles, S., & Kosack, G. (1973). *Immigrant workers and class structure in western Europe*. Oxford University Press.

Catney, G. (2018). The complex geographies of ethnic residential segregation: Using spatial and local measures to explore scale-dependency and spatial relationships. *Transactions of the Institute of British Geographers, 43*(1), 137–152.

Cesari, J. (2005). Mosque conflicts in European cities: Introduction. *Journal of Ethnic and Migration Studies, 31*(6), 1015–1024.

Chiodelli, F. (2015). Religion and the city: A review on Muslim spatiality in Italian cities. *Cities, 44*, 19–28.

Christiansen, L. B., Galal, L. P., & Hvenegaard-Lassen, K. (2017). Organised cultural encounters: Interculturality and transformative practices. *Journal of Intercultural Studies, 38*(6), 599–605.

Cocola-Gant, A., & Lopez-Gay, A. (2020). Transnational gentrification, tourism and the formation of "foreign only" enclaves in Barcelona. *Urban Studies, 57*(15), 3025–3043.

Colombo, E. & Semi, G. (Eds.). (2007). *Multiculturalismo quotidiano: Le pratiche della differenza* (Everyday multiculturalism: The practices of difference). FrancoAngeli.

Conradson, D., & Latham, A. (2005). Transnational urbanism: Attending to everyday practices and mobilities. *Journal of Ethnic and Migration Studies, 31*(2), 227–233.

Coulter, R., van Ham, M., & Findlay, A. M. (2016). Re-thinking residential mobility: Linking lives through time and space. *Progress in Human Geography, 40*(3), 352–374.

Cressey, P. (1932). *The taxi-dance hall: A sociological study in commercialized recreation and city life*. University of Chicago Press.

Dahya, B. (1974). The nature of Pakistani ethnicity in industrial cities in Britain. In A. Cohen (Ed.), *Urban ethnicity* (pp. 77–118). Tavistock.

Darling, J., & Wilson, H. F. (2016). *Encountering the city: Urban encounters from Accra to New York*. Routledge.

Dawes, A. L. (2020). *Race talk: Language of racism and resistance in Neapolitan street markets*. Manchester University Press.

Della Puppa, F., & King, R. (2019). The new "twice migrants": Motivations, experiences and disillusionments of Italian-Bangladeshis relocating to London. *Journal of Ethnic and Migration Studies, 45*(11), 1936–1952.

De Maillard, J., Verfaillie, K., & Rowe, M. (2024). *The politicization of police stops in Europe. Public issues and police reform*. Palgrave.

De Noronha, L. (2022). The conviviality of the overpoliced, detained and expelled: Refusing race and salvaging the human at the borders of Britain. *The Sociological Review, 70*(1), 159–177.

DeVerteuil, G., & Manley, D. (2017). Overseas investment into London: Imprint, impact and pied-à-terre urbanism. *Environment and Planning A: Economy and Space, 49*(6), 1308–1323.

Díaz-Cortés, F., Albet-Mas, A., & Garcia-Ramon, M. D. (2008). Old and new migrant women in Ca N'Anglada: Public spaces, identity and everyday life in the metropolitan region of Barcelona. In J. N. DeSena (Ed.), *Gender in an urban world* (pp. 263–284). Emerald Group Publishing.

Dietrich-Ragon, P. (2017). On the sidelines of French society: Homelessness among migrants and their descendants. *Population, 72*(1), 7–38.

Dines, N. (2002). Urban renewal, immigration and contested claims to public space: The case of Piazza Garibaldi in Naples. *GeoJournal, 58*(2–3), 177–188.

Dines, N. (2009). The disputed place of ethnic diversity: An ethnography of the redevelopment of a street market in East London. In R. Imrie, L. Lees, & M. Raco (Eds.), *Regenerating London. Governance, sustainability and community in a global city* (pp. 254–272). Routledge.

Dines, N. (2012). *Tuff city: Urban change and contested space in central Naples*. Berghahn.

Dines, N., & Mattiucci, C. (2022). Constantly evoked but under-researched: The conundrum of vertical stratification in Naples. In T. Maloutas & N. Karadimitriou (Eds.), *Vertical Cities: Micro-segregation, social mix and urban housing markets* (pp. 23–38). Elgar.

Domínguez-Mujica, J., McGarrigle, J., & Parreño-Castellano, J. M. (Eds.). (2021). *International residential mobilities: From lifestyle migrations to tourism gentrification*. Springer.

Dotsey, S., & Ambrosini, M. (2025). Migration, (il)legal status and housing precarity: Difficulties and coping strategies. *Journal of Urban Affairs, 47*(4), 1328–1347.

Dotsey, S., & Chiodelli, F. (2021). Housing precarity: A fourfold epistemological lancet for dissecting the housing conditions of migrants. *City, 25*(5–6), 720–739.

Dwyer, C., Gilbert, D., & Shah, B. (2013). Faith and suburbia: Secularisation, modernity and the changing geographies of religion in London's suburbs. *Transactions of the Institute of British Geographers, 38*(3), 403–419.

Eade, J. (2017). Creating religious homes in London: Sacralising space in a deeply globalised city. In F. Guesnet, C. Laborde, & L. Lee (Eds.), *Negotiating religion: Cross-disciplinary perspectives* (pp. 151–168). Routledge.

Ehrkamp, P. (2005). Placing identities: Transnational practices and local attachments of Turkish immigrants in Germany. *Journal of Ethnic and Migration Studies, 31*(2), 345–364.

Ehrkamp, P. (2008). Risking publicity: Masculinities and the racialization of public neighbourhood space. *Social & Cultural Geography, 9*(2), 117–133.

Eskelä, E. (2018). Housing pathways of skilled migrants: Indian professionals in Helsinki, Finland. *Housing, Theory and Society, 35*(4), 474–494.

Fallov, M. A., & Birk, R. H. (2022). The "Ghetto" strikes back: Resisting welfare sanctions and stigmatizing categorizations in marginalized residential areas in Denmark. *Nordic Social Work Research, 12*(2), 217–228.

Fathi, M., & Ní Laoire, C. (2023). Urban home: Young male migrants constructing home in the city. *Journal of Ethnic and Migration Studies, 49*(3), 813–831.

Fathi, M., & Ní Laoire, C. (2024). *Migration and home.* Springer.

Fauser, M. (2019). The emergence of urban border spaces in Europe. *Journal of Borderlands Studies, 34*(4), 605–622.

Fernandez, R., Hofman, A., & Aalbers, M. (2016). London and New York as a safe deposit box for the transnational wealth elite. *Environment and Planning A: Economy and Space, 48*(12), 2443–2461.

Finney, N., & Simpson, L. (2009). *Sleepwalking to segregation: Challenging myths about race and migration.* Policy Press.

Foot, J. (2001). *Milan since the miracle: City, culture and identity.* Berg.

Gale, R. (2004). The multicultural city and the politics of religious architecture: Urban planning, mosques and meaning-making in Birmingham UK. *Built Environment, 30*(1), 30–44.

Garbin, D., & Strhan, A. (Eds.). (2017). *Religion and the global city.* Bloomsbury.

Garnett, J., & Harris, A. (Eds.). (2013). *Rescripting religion in the city. Migration and religious identity in the modern metropolis.* Routledge.

Gawlewicz, A., & Yiftachel, O. (2022). "Throwntogetherness" in hostile environments. Migration and the remaking of urban citizenship. *City, 26*(2–3), 346–358.

Gielis, R. (2009). A global sense of migrant places: Towards a place perspective in the study of migrant transnationalism. *Global Networks, 9*(2), 271–287.

Gilbert, D., Dwyer, C., Ahmed, N., Cuch, L., & Hyacinth, N. (2019). The hidden geographies of religious creativity: Place-making and material culture in west London faith communities. *Cultural Geographies, 26*(1), 23–41.

Gilroy, P. (2004). *After empire: Melancholia or convivial culture?* Routledge.

Glick Schiller, N., Çağlar, A., & Guldbrandsen, T. (2006). Beyond the ethnic lens: Locality, globality, and born-again incorporation. *American Ethnologist, 33*(4), 612–633.

Glynn, S. (2010). Playing the ethnic card: Politics and segregation in London's East End. *Urban Studies, 47*(5), 991–1013.

Goffman, E. (1963). *Behavior in public places: Notes on the social organization of gatherings.* Free Press.

Göle, N. (2016). *Islam and public controversy in Europe.* Ashgate.

Grillo, R. (1985). *Ideologies and institutions in urban France: The representation of immigrants.* Cambridge University Press.

Gruner, S. (2010). "The others don't want…" Small-scale segregation: Hegemonic public discourses and racial boundaries in German neighbourhoods. *Journal of Ethnic and Migration Studies, 36*(2), 275–292.

Gustafson, P. (2009). Your home in Spain: Residential strategies in international retirement migration. In M. Benson & K. O'Reilly (Eds.), *Lifestyle migration: Expectations, aspirations and experiences* (pp. 69–86). Ashgate.

Hall, S. (1993). Culture, community, nation. *Cultural Studies, 7*(3), 349–363.

Hall, S. M. (2015). Migrant urbanisms: Ordinary cities and everyday resistance. *Sociology, 45*(5), 853–869.

Hamann, U., & Türkmen, C. (2020). Communities of struggle the making of a protest movement around housing migration and racism beyond identity politics in Berlin. *Territory, Politics, Governance, 8*(4), 515–531.

Hardy, S. (2017). *Everyday multiculturalism and "hidden" hate.* Palgrave.

Hentschel, C., & Blokland, T. (2016). Life and death of the great regeneration vision: Diversity, decay, and upgrading in Berlin's ordinary shopping streets. In S. Zukin, P. Kasinitz, & X. Chen (Eds.), *Global cities, local streets: Everyday diversity from New York to Shanghai* (pp. 120–139). Routledge.

Hewitt, P. (1986). *White talk black talk: Inter-racial friendship and communication among adolescents.* Cambridge University Press.

Iveson, K. (2007). *Publics and the city.* Blackwell.

Jackson, E. (2019). Valuing the bowling alley: Contestations over the preservations of spaces of everyday urban multiculture in London. *The Sociological Review, 67*(1), 79–94.

Janssens, J., & Verweel, P. (2014). The significance of sports clubs within multicultural society. On the accumulation of social capital by migrants in culturally "mixed" and "separate" sports clubs. *European Journal for Sport and Society, 11*(1), 35–58.

Jessup, S., & Jackson, L. (2024, February 26). Paul Scully MP defends saying parts of Tower Hamlets "no-go" areas. *BBC London News.* https://www.bbc.com/news/uk-england-london-68407084

Jones, H., Gunaratnam, Y., Bhattacharyya, G., Davies, W., Dhaliwal, S., Forkert, K., Jackson, E., & Saltus, R. (2017). *Go home? The politics of immigration controversies.* Manchester University Press.

Jung, P. R., & Buhr, F. (2022). Channelling mobilities: Migrant-owned businesses as mobility infrastructures. *Mobilities, 17*(1), 119–135.

Karner, C., & Parker, D. (2011). Conviviality and conflict: Pluralism, resilience and hope in inner-city Birmingham. *Journal of Ethnic and Migration Studies, 37*(3), 355–372.

Keskiner, E., & Waldring, I. (2023). Revisiting the established-outsider constellation in a gentrifying majority-minority neighbourhood. *Journal of Ethnic and Migration Studies, 49*(8), 2052–2069.

Knights, M., & King, R. (1998). The geography of Bangladeshi migration to Rome. *International Journal of Population Geography, 4*, 299–321.

Kohlenberger, J. (2024). Social infrastructure for migrants and integration. In A. Renner, L. Plank, & M. Getzner (Eds.), *Handbook of social infrastructure: Conceptual and empirical research perspectives* (pp. 160–175). Elgar.

Koulaxi, A. M. (2022). Convivial reflexivity in the changing city—A tale of hospitality or hostility? *International Journal of Cultural Studies, 25*(2), 210–230.

Kreichauf, R. (2018). From forced migration to forced arrival: the campization of refugee accommodation in European cities. *Comparative Migration Studies, 6*(7), https://doi.org/10.1186/s40878-017-0069-8

Krifors, K. (2022). Rural multiculturalism? Migrants, antiracism, and convivial cultures in provincial Sweden. *Ethnic and Racial Studies, 45*(16), 72–92.

Lapina, L. (2016). Besides conviviality: Paradoxes in being "at ease" with diversity in a Copenhagen district. *Nordic Journal of Migration Research, 6*(1), 33–41.

Latham, A., & Layton, J. (2019). Social infrastructure and the public life of cities: Studying urban sociality and public spaces. *Geography Compass, 13*, 1–15.

Lauwers, A. S. (2023). Religion, secularity, culture? Investigating Christian privilege in Western Europe. *Ethnicities, 23*(3), 403–425.

Leal, J. (2004). Segregation and social change in Madrid metropolitan region. *The Greek Review of Social Research, 113*, 81–104.

Lofland, L. H. (1998). *The public realm: Exploring the city's quintessential social territory*. Aldine de Gruyter.

Lombard, M. (2023). The experience of precarity: Low-paid economic migrants' housing in Manchester. *Housing Studies, 38*(2), 307–326.

Lukas T. (2009). Why are there no riots in Germany? Mutual perceptions between the police and minority adolescents. In D. Waddington, F. Jobard, & M. King (Eds.), *Rioting in the UK and France: A comparative analysis* (pp. 216–28). Willan.

Lundström, C. (2014). *White migrations: Gender, whiteness and privilege in transnational migration*. Palgrave Macmillan.

Malheiros, J. (2002). Ethni-cities: Residential patterns in the northern European and Mediterranean metropolises—Implications for policy design. *International Journal of Population Geography, 8*, 107–134.

Maloutas, T. (2007). Segregation, social polarization and immigration in Athens during the 1990s: Theoretical expectations and contextual difference. *International Journal of Urban and Regional Research, 31*(4), 733–758.

Maloutas, T., & Spyrellis, S. N. (2016). Vertical segregation: Mapping the vertical social stratification of residents in Athenian apartment buildings. *Méditerranée, 127*, 27–36.

Manting, D., Kleinepier, T., & Lennartz, C. (2024). Housing trajectories of EU migrants: Between quick emigration and shared housing as temporary and long-term solutions. *Housing Studies, 39*(4), 1027–1048.

Manzano Gómez, N. A. (2024). Interrogating the origins of informal urbanisation: A socio-historical analysis from Paris and Madrid (1850s–1970s). *European Urban and Regional Studies, 31*(4), 346–359.

Marabello, S., & Riccio, B. (2020). Spazi di convivialità? Convivere e co-abitare con migrant in Italia: Introduzione (Spaces of conviviality? Living and cohabiting with migrants in Italy: Introduction). *Antropologia Pubblica, 6*(2), 25–32.

Marcińczak, S., Tammaru, T., Strömgren, M., & Lindgren, U. (2015). Changing patterns of residential and workplace segregation in the Stockholm metropolitan area. *Urban Geography, 36*(15), 969–992.

Martikainen, T., Mapril, J., & Khan, A. H. (2019). *Muslims at the margins of Europe. Finland, Greece, Ireland and Portugal*. Brill.

Maslova, S., & King, R. (2020). Residential trajectories of high-skilled transnational migrants in a global city: Exploring the housing choices of Russian and Italian professionals in London. *Cities, 96*. https://doi.org/10.1016/j.cities.2019.102421

McGarrigle, J. (2016). Islam in urban spaces: The residential incorporation and choices of Muslims in Lisbon. *Journal of Ethnic and Migration Studies, 42*(3), 437–457.

McGarrigle, J., & Ascensão, E. (2018). Emplaced mobilities: Lisbon as a translocality in the migration journeys of Punjabi Sikhs to Europe. *Journal of Ethnic and Migration Studies, 44*(5), 809–828.

McKee, K., Leahy, S., Tokarczyk, T., & Crawford, J. (2021). Redrawing the border through the "Right to Rent": Exclusion, discrimination and hostility in the English housing market. *Critical Social Policy, 41*(1), 91–110.

Meeus, B. (2017). Challenging the figure of the "migrant entrepreneur". Place-based solidarities in the Romanian arrival infrastructure in Brussels. In S. Oosterlynck, N. Schuermans, & M. Loopmans (Eds.), *Place, diversity and solidarity* (pp. 91–108). Routledge.

Meeus, B., Arnaut, K., & van Heur, B. (2019). *Arrival infrastructures: Migration and urban social mobilities*. Palgrave Macmillan.

Mitchell, K. (2011). Marseille's not for burning: Comparative networks of integration and exclusion in two French cities. *Annals of the Association of American Geographers, 101*(2), 404–423.

Montagna, N., & Grazioli, M. (2019). Urban commons and freedom of movement: The housing struggles of recently arrived migrants in Rome. *Citizenship Studies, 23*(6), 577–592.

Montezuma, J., & McGarrigle, J. (2019). What motivates international homebuyers? Investor to lifestyle "migrants" in a tourist city. *Tourism Geographies, 21*(2), 214–234.

Morad, M., Della Puppa, F., & Sacchetto, D. (2021). The dark side of onward migration: Experiences and strategies of Italian-Bangladeshis in the UK at the time of the post-Brexit referendum. *British Journal of Sociology, 7*(5), 1311–1324.
Mudu, P., & Chattopadhyay, S. (Eds.). (2017). *Migration, squatting and radical autonomy*. Routledge.
Murdie, R. A., & Borgegård, L. (1998). Immigration, spatial segregation and housing segmentation of immigrants in Metropolitan Stockholm, 1960–95. *Urban Studies, 35*(10), 1869–1888.
Musterd, S. (2023). *Advanced introduction to urban segregation*. Elgar.
Musterd, S., & Ostendorf, W. (1998). *Urban segregation and the welfare state: Inequality and exclusion in western cities*. Routledge.
Musterd, S., Ostendorf, W., & Breebaart, M. (1998). *Multi-ethnic metropolis: Patterns and policies*. Springer.
Najib, K., & Hopkins, P. (2020). Where does Islamophobia take place and who is involved? Reflections from Paris and London. *Social & Cultural Geography, 21*(4), 458–478.
Narkowicz, K., & Pędziwiatr, K. (2017). From unproblematic to contentious: Mosques in Poland. *Journal of Ethnic and Migration Studies, 43*(3), 441–457.
Nayak, A. (2010). Race, affect, and emotion: Young people, racism, and graffiti in the postcolonial English suburbs. *Environment and Planning A: Economy and Space, 42*(10), 2370–2392.
Nayak, A. (2017). Purging the nation: Race, conviviality and embodied encounters in the lives of British Muslim young women. *Transactions of the Institute of British Geographers, 42*(2), 289–302.
Neal, S., Bennett, K., Cochrane, A., & Mohan, G. (2013). Living multiculture: Understanding the new spatial and social relations of ethnicity and multiculture in England. *Environment and Planning C: Government and Policy, 31*(2), 308–323.
Newburn, T. (2021). The causes and consequences of urban riot and unrest. *Annual Review of Criminology, 4*, 53–73.
Noussia, A., & Lyons, M. (2009). Inhabiting spaces of liminality: Migrants in Omonia, Athens. *Journal of Ethnic and Migration Studies, 35*(4), 601–624.
Obradovic-Wochnik, J. (2018). Urban geographies of refugee journeys: Biopolitics, neoliberalism and contestation over public space in Belgrade. *Political Geography, 67*, 65–75.
Oosterbaan, M. (2014). Public religion and urban space in Europe. *Social & Cultural Geography, 15*(6), 591–602.
Pastore, F., & Ponzo, I. (2016). *Inter-group relations and migrant integration in European cities. Changing neighbourhoods*. Springer.
Padilla, B., Azevedo, J., & Olmos-Alcaraz, A. (2015). Superdiversity and conviviality: Exploring frameworks for doing ethnography in Southern European intercultural cities. *Ethnic and Racial Studies, 38*(4), 621–635.
Peach, C. (1996). Does Britain have ghettos? *Transactions of the Institute of British Geographers, 21*(1), 216–235.
Peach, C. (1999). London and New York: Contrasts in British and American models of segregation with a comment by Nathan Glazer. *International Journal of Population Geography*, *5*(5), 319–351.
Peach, C. (1998). South Asian and Caribbean ethnic minority housing choice in Britain. *Urban Studies, 35*(10), 1657–1680.
Peach, C. (2005, September 28). Britain has enclaves, not ghettoes. *The Guardian*. https://www.theguardian.com/uk/2005/sep/28/race.world
Peach, C. (2009). Slippery segregation: Discovering or manufacturing ghettos? *Journal of Ethnic and Migration Studies, 35*(9), 1381–1395.
Pemberton, S., & Phillimore, J. (2018). Migrant place-making in super-diverse neighbourhoods: Moving beyond ethno-national approaches. *Urban Studies, 55*(4), 733–750.
Picker, G. (2017). *Racial cities: Governance and the segregation of Romani people in urban Europe*. Routledge.

Piekut, A. (2021). Re-theorising spatial segregation: A European perspective. In G. Pryce, Y. Ping, Y. Chen, J. Shan, & H. Wei (Eds.), *Urban inequality and segregation in Europe and China: Towards a new dialogue* (pp. 13–38). Springer.

Phillips, D. (2006). Parallel lives? Challenging discourses of British Muslim self-segregation. *Environment and Planning D: Society and Space, 24*(1), 25–40.

Phillips, D. (2007). Ethnic and racial segregation: A critical perspective. *Geography Compass, 1*(5), 1138–1159.

Phillips, D. (2015). "Race", community and ongoing conflict. *Ethnic and Racial Studies, 38*(3), 391–397.

Phillips, D., Davis, C., & Ratcliffe, P. (2007). British Asian narratives of urban space. *Transactions of the Institute of British Geographers, 32*(2), 217–234.

Pompeo, F. (2011) *Pigneto-Banglatown. Migrazioni e conflitti di cittadinanza in una periferia storica romana* (Pigneto-Banglatown. Migration and conflicts among the citizenry in a historic Roman periphery). Meti Edizioni.

Raco, M., & Kesten, J. (2018). The politicisation of diversity planning in a global city: Lessons from London. *Urban Studies, 55*(4), 891–916.

Ramamurthy, A. (2013). *Black star: Britain's Asian youth movements.* Pluto Press.

Rangvid, B. S. (2007). Living and learning separately? Ethnic segregation of school children in Copenhagen. *Urban Studies, 44*(7), 1329–1354.

Rex, J., & Moore, R. (1967). *Race, community and conflict: A study of Sparkbrook.* Oxford University Press.

Risager, B. S. (2023). Territorial stigmatization and housing commodification under racial neoliberalism: The case of Denmark's "ghettos." *Environment and Planning A: Economy and Space, 55*(4), 850–870.

Rishbeth, C., & Powell, M. (2013). Place attachment and memory: Landscapes of belonging as experienced post-migration. *Landscape Research, 38*(2), 160–178.

Rishbeth, C., Blachnicka-Ciacek, D., & Darling, J. (2019). Participation and wellbeing in urban greenspace: "Curating sociability" for refugees and asylum seekers. *Geoforum, 106*, 125–134.

Robinson, D., Reeve K., & Casey, R. (2007). *The housing pathways of new immigrants.* Joseph Rowntree Foundation.

Robinson, K. (2020). Everyday multiculturalism in the public library: Taking knitting together seriously. *Sociology, 54*(3), 556–572.

Rogaly, B. (2020). *Stories from a migrant city: Living and working together in the shadow of Brexit.* Manchester University Press.

Rosbrook-Thompson, J. (2018). Understanding difference amid superdiversity: Space, "race" and granular essentialisms at an inner-city football club. *Sociology, 52*(4), 639–654.

Ryan, L. (2018). Differentiated embedding: Polish migrants in London negotiating belonging over time. *Journal of Ethnic and Migration Studies, 44*(2), 233–251.

Saifullah Khan, V. (1976). Pakistanis in Britain: Perceptions of a population. *New Community, 5*(3), 222–229.

Saint-Blancat, C., & Schmidt di Friedberg, O. (2005). Why are mosques a problem? Local politics and fear of Islam in Northern Italy. *Journal of Ethnic and Migration Studies, 31*(6), 1083–1104.

Samanani, F. (2023). Conviviality and its others: For a plural politics of living with difference. *Journal of Ethnic and Migration Studies, 49*(9), 2109–2128.

Santoro, G. (2015). *Al palo della morte. Storia di un omicidio in una periferia meticcia* (At the pole of death: Story of a murder in a mixed suburb). Edizioni Alegre.

Schier, M. (2016). Everyday practices of living in multiple places and mobilities: Transnational, transregional, and intra-communal multi-local families. In M. Kiley & E. Palenga-Möllenbeck (Eds.), *Family life in an age of migration and mobility: Global perspectives through the life course* (pp. 43–69). Palgrave Macmillan.

Schierup, C., Ålund, A., & Kings, L. (2014). Reading the Stockholm riots—A moment for social justice? *Race & Class, 55*(3), 1–21.

Schiller, M., & Raco, M. (2021). Postcolonial narratives and the governance of informal housing in London. *International Journal of Housing Policy, 21*(2), 268–290.
Schmoll, C. (2019). A city of one's own. Territorialities of migrant women in public spaces. In I. Micha & D. Vaiou (Eds.), *Alternative takes to the city* (pp. 1–20). ISTE.
Simmel, G. (1950 [1908]). The stranger. In K. H. Wolff (Ed.), *The sociology of Georg Simmel* (pp. 402–408). Free Press.
Slater, T., & Anderson, N. (2012). The reputational ghetto: Territorial stigmatisation in St Paul's, Bristol. *Transactions of the Institute of British Geographers, 37*(4), 530–546.
Slater, T. (2021). *Shaking up the city: Ignorance, inequality, and the urban question.* University of California Press.
Spena, A. (2016). Resisting immigration detention. *European Journal of Migration Law, 18*, 201–221.
Stanfield, E., & van Riemsdijk, M. (2019). Creating public space, creating "the public": Immigration politics and representation in two Copenhagen parks. *Urban Geography, 40*(9), 1356–1374.
Stansfeld, K. (2024). "Are we invisible?" Power-geometries of conviviality in a superdiverse London neighbourhood. *Transactions of the Institute of British Geographers, 49*(4), 1–16.
Stehle, M. (2006). Narrating the ghetto, narrating Europe: From Berlin, Kreuzberg to the *banlieues* of Paris. *Westminster Papers in Communication and Culture, 3*(3), 48–70.
Stonawski, M., Brzozowski, J., Pędziwiatr, K., & Georgati, M. (2022). Investigating neighbourhood concentration of immigrants in Poland: Explorative evidence from Kraków. *Bulletin of Geography. Socio-Economic Series, 56*, 143–159.
Sýkora, L. (2009). New socio-spatial formations: Place of residential segregation and separation in Czechia. *Tijdschrift Voor Economische En Sociale Geografie, 100*(4), 417–435.
Tammaru, T., Marcińczak, S., van Ham, M., & Musterd, S. (2016). *Socio-economic segregation in European capital cities: East meets West*. Routledge.
Tasan-Kok, T., van Kempen, R., Raco, M., & Bolt, G. (2014). *Towards hyper-diversified European cities: A critical literature review*. CPI Wohrmann.
Torpan, K., Sinitsyna, A., Kährik, A., Kauppinen, T. M., & Tammaru, T. (2022). Overlap of migrants' housing and neighbourhood mobility. *Housing Studies, 37*(8), 1396–1421.
Tyler, K. (2017). The suburban paradox of conviviality and racism in postcolonial Britain. *Journal of Ethnic and Migration Studies, 43*(11), 1890–1906.
Valentine, G. (2008). Living with difference: Reflections on geographies of encounter. *Progress in Human Geography, 32*(3), 323–337.
Valluvan, S. (2016). Conviviality and multiculture: A post-integration sociology of multi-ethnic interaction. *Young, 24*(3), 204–221.
van der Horst, H. (2010). Dwellings in transnational lives: A biographical perspective on "Turkish-Dutch" houses in Turkey. *Journal of Ethnic and Migration Studies, 36*(7), 1175–1192.
van Kempen, R. & Özüekren, A. S. (1998). Ethnic segregation in cities: New forms and explanations in a dynamic world. *Urban Studies, 35*(10), 1631–1656.
van Kempen, R., & van Weesep, J. (1998). Ethnic residential patterns in Dutch cities: Backgrounds, shifts and consequences. *Urban Studies, 35*(10), 1813–1833.
Venkov, N. A. (2019). Conviviality vs politics of coexistence: Going beyond the global North. *CAS Sofia Working Paper Series, 11*, 1–42.
Verhaeghe, P. P., & De Coninck, D. (2022). Rental discrimination, perceived threat and public attitudes towards immigrants and refugees. *Ethnic and Racial Studies, 45*(7), 1371–1393.
Vertovec, S. (2007). Super-diversity and its implications. *Ethnic and Racial Studies, 30*(6), 1024–1054.
Vincent, C., Neal, S., & Iqbal, H. (2018). Living in the city: School friendships, diversity and the middle classes. *The British Journal of Sociology, 69*(2), 352–371.
Vine, J. (2024). Chronic urban trauma, postsecular spaces and living with difference in a city neighbourhood. *Migration Studies, 12*(2). https://doi.org/10.1093/migration/mnad036
Wacquant, L. (2007). *Urban outcasts. A comparative sociology of advanced marginality.* Polity Press.

Waddington, D., Jobard, F., & King, M. (Eds.). (2009). *Rioting in the UK and France: A comparative analysis*. Willan.
Wessendorf, S. (2014). *Commonplace diversity. Social relations in a super-diverse context*. Palgrave Macmillan.
Wessendorf, S. (2018). Pathways of settlement among pioneer migrants in super-diverse London. *Journal of Ethnic and Migration Studies, 44*(2), 270–286.
Wessendorf, S., & Gembus, M. (2024). The social front door: The role of social infrastructures for migrant arrival. *Journal of Ethnic and Migration Studies, 50*(12), 2822–2838.
White, P. (1999). Minority residential histories in the city: Contexts, processes and outcomes. In Fundação Luso-Americana para o Desenvolvimento (Ed.), *Metropolis international workshop proceedings* (pp. 155–186). FLAD.
White, P., & Hurdley, L. (2003). International migration and the housing market: Japanese corporate movers in London. *Urban Studies, 40*(4), 687–706.
Wilkins, A. (2019). *Migration, work and home-making in the city: Dwelling and belonging among Vietnamese communities in London*. Routledge.
Wilson, H. F. (2011). Passing propinquities in the multicultural city: The everyday encounters of bus passengering. *Environment and Planning A: Economy and Space, 43*(3), 634–649.
Wilson, H. F. (2017). On geography and encounter: Bodies, borders, and difference. *Progress in Human Geography, 41*(4), 451–471.
Wimmer, A. (2004). Does ethnicity matter? Everyday group formation in three Swiss immigrant neighbourhoods. *Ethnic and Racial Studies, 27*(1), 1–36.
Wise, A., & Noble, G. (2016). Convivialities: An orientation. *Journal of Intercultural Studies, 37*(5), 423–431.
Wise, A., & Velayutham, S. (2009). *Everyday multiculturalism*. Palgrave Macmillan.
Zukin, S., Kasinitz, P., & Chen, X. (2016). *Global cities, local streets: Everyday diversity from New York to Shanghai*. Routledge.

Chapter 5
Migration and Urban Cultures: Representations, Practices and Consumption

5.1 Introduction

It woz di miggle a di rush howah
wen evrybady jus a hosel an a bosel
fi go home fi dem evenin showah;
mi an Jim stan-up
waiting pan a bus,
nat cauzin no fus,
wen all af a sudden
a police van pull-up (Johnson, 1991: 25)[1]

This verse is taken from 'Sonny's Lettah (Anti-Sus Poem)', written and performed by Linton Kwesi Johnson for his 1979 dub poetry album *Forces of Victory*. Born in Jamaica in 1952, Johnson moved to England at the age of eleven to live with his mother in the London district of Brixton. 'Sonny's Lettah' combines an account of everyday life for Caribbean youth on the streets of south London with a denunciation of racism and police harassment. It epitomizes Johnson's pioneering use of London-Jamaican patois. The English language, he explained in a film interview, was 'too dead, too sterile to communicate the violence of [things which were happening around me].'[2] Although rooted in Jamaican toasting, dub poetry was as much a creation of Brixton, thanks to the area's concentration of Caribbean musicians, all-night reggae clubs, portable sound systems and record shops. The Kittitian-British novelist Caryl Phillips described Linton Kwesi Johnson as 'the first crossover voice, who made it possible for a generation to think of themselves as black and creative in literature,

[1] 'It was in the middle of the rush hour/ when everyone is just hustling and bustling/ to get home for their evening showers/ me and Jim were standing/ waiting for a bus/ not causing any fuss/ when all of a sudden/ a police van pulled up.' The song version can be heard here: https://www.youtube.com/watch?v=7uvY5qU7ayg.

[2] Citation from an interview in Franco Rossi's 1979 documentary film *Dread Beat and Blood*, an extract of which can be viewed here: https://www.youtube.com/watch?v=YZoO2Iz9jLw.

N. Dines, *Migration and European Cities*, IMISCOE Research Series,
https://doi.org/10.1007/978-3-032-15379-1_5

music, the media' (Jaggi, 2002), while dub music would have a major influence on other urban genres such as hip hop and punk. Johnson would develop an international following and tour extensively in Europe and North America. Meanwhile, Brixton has remained an iconic centre of black popular culture in the British capital, although it has long been a place of arrival and settlement for other migrant groups and has often been stigmatized for its socio-economic problems and conflicts, most notably the 1981 Brixton riots. Since the 1990s, Brixton has also become increasingly gentrified as mainly white professionals and creative workers have moved into the area, attracted by its proximity to central London, its Victorian properties, its vibrant night-time economy and a multicultural vibe in which dub reggae, divorced from its social and political moorings, provides a congenial backdrop (Lees, 2016).

The following passage is taken from the opening voiceover of *Bangla*, a 2019 romantic comedy by Italian-Bangladeshi director Phaim Bhuiyan. Compared to 'Sonny's Lettah', it marks a decisive shift in tone.

> Even if I look like a black man to you, I'm actually Italian…well, sort of half way, like a cappuccino: 50% Bangla, 50% Italian, 100% Torpigna. Torpignattara, popularly known as Torpigna, is the most multi-ethnic neighbourhood of Rome. Since Piazza Vittorio's become all fancy, we're the new frontier zone. I was born here a couple of years after my parents arrived in Italy. Torpigna has recently gone all trendy, especially when it comes to black people, immigration, terrorist attacks and what not. Truth be told, there's a shitload of different people here; a kind of globalization. You walk around here and at any time of the day you can smell lasagne, curry, kebabs – all mixed up – it's a psychedelic experience.[3]

Bangla tells the story of 22-year-old Phaim, played and narrated by Bhuiyan himself, who falls in love with Asia, a young white Italian woman from an upper-middle-class neighbourhood in northern Rome. A minor box office success, the film went on to win several Italian awards and was adapted into a Netflix spin-off series. Drawing on the director's own experience of growing up in Torpignattara, *Bangla* uses the distinctive slang and humour of Rome to play with stereotypes about gender, family and sexuality among people with migrant backgrounds, and to poke fun at sociological clichés about the Italian capital. Adorned with street art, Torpignattara doubles up as the ramshackle backdrop in which Phaim and his friends live out their humdrum lives, from rehearsing with their Bangladeshi pop band to seeking advice from unlikely peers at the local mosque and park. The city centre is a nondescript place of dead-end service jobs, the bourgeois milieu of Roma Nord, home to Asia and her preposterously progressive parents, is cast as a foreign land, while London appears on the horizon as a destination that can fulfil Bangladeshi aspirations for upward mobility (although Phaim's companion, who manages to relocate there, soon finds himself longing for the food and social life of Rome). Alongside the fraught love story, the film shines light on local attachments to place, everyday cultural practices and translocal connections as well as the ubiquitous presence of social media and digital communication. As a commercial film, *Bangla* was inevitably conditioned by the demands of the Italian film industry and while it was praised in the mainstream press, it was read in more critical quarters as a place-branding exercise that sought to

---

[3] The opening to *Bangla* can be viewed here: https://www.youtube.com/watch?v=m6opq5inI-w.

reassure the public through its safe portrayal of 'good' Italians (Caroselli et al., 2019). Nevertheless, through its light-hearted and irreverent register, the film breaks new ground in its depiction of urban diversity in Italy, while capturing the ideological impact of neoliberal individualism on the livelihoods of young people of migrant origin in the contemporary city.

'Sonny's Lettah' and *Bangla* are just two examples of the artistic activities and production methods, as well as the creative industries and audiences, that emerge from the intersections of migration and cities. On the one hand, London/Brixton and Rome/Torpignattara function as sites and subjects of representation; on the other, migration and its aftermaths structure the creation of new place-based meanings and relationships. Linton Kwesi Johnson's dub poetry and Phaim Bhuiyan's film are historically and geographically situated: both draw on and feed into the cultural landscapes of particular places at particular moments in time. Listening to the thick Roman accents of Bangladeshi youths in *Bangla* has a striking effect given the city's relatively short history of large-scale international migration and the general lack of non-white voices in public life. Hearing London patois today has a different impact in the British capital, where music and street language have long represented zones of exchange between black, Asian and white youth (James, 2015). Meanwhile, the creative works of Johnson and Bhuiyan resonate with different local, national and international publics through their respective circuits of distribution, potentially altering perceptions of cities and offering modes of expression that can be reproduced and modified elsewhere.

'Culture' is a notoriously elusive concept. A distinction is conventionally drawn between culture as the product of human creativity and culture as a 'way of life', i.e. the practices, behaviours, knowledge, values and institutions that are seen to characterize social groups. There has also been a traditional division between 'high' and 'popular' culture—the former identified with the ruling class and ideas of civilization, and the latter with mass society—both of which were closely associated with urban life from the nineteenth century onwards (Williams, 1983). Any attempt at an all-encompassing definition of culture is further complicated when we consider migration. Since the 1980s various theorists such as Arjun Appadurai (1996), Homi Bhabha (1994), Avtar Brah (1996), Paul Gilroy (1993) and Stuart Hall (2021 [2006]) have explored in different ways how human mobility, transnational networks and globalization alongside the politics of colonialism and racism, undermine any claim to a discrete and static notion of culture. Instead, processes of hybridization, creolization, in-betweenness, contamination and resistance have given rise, especially in large cities, to a panoply of syncretic styles, diasporic configurations and 'ethnoscapes' (Appadurai, 1996). On the other hand, there is a persistent tendency in popular discourse to perceive people with migrant backgrounds as belonging to distinct cultural communities, which can operate to reify group differences and, especially in the case of racialized minorities, reaffirm their subordinate position in society (Çağlar, 1997).

This chapter does not attempt to provide a theoretical model or inventory of the cultural practices and lifestyles associated with migration in urban settings. The myriad permutations render such exercises somewhat futile. Rather, the main aim here is to place cultural processes at the centre of our understanding of the migration-city nexus and to think through the consequences of adopting such a perspective. Examining the situated nature of cultural production can offer more nuanced insights into how people view their own or others' migration, while shedding light on the ways in which they respond to and shape the urban realities around them. In doing so, it can more fully address issues of subjective experience, public recognition and meaning-making that have been marginal to the study of migration and cities.

Indeed, until recently, migration studies tended to overlook, or simply ignore, the cultural practices of people with migrant backgrounds. This was partly because, as Marco Martiniello argues, 'immigrants and their offspring have for a long time been exclusively considered as workers, as mere means of production in the industrial economy or in the post-industrial service economy' (Martiniello, 2015: 1230). The creative activities of migrants were typically the domain of other sub-disciplines such as ethnomusicology, film studies or, particularly in the case of the UK, cultural studies. All this may seem at odds with the fact that academic and policy discourses on migration and cities have habitually used terms such as 'cultural diversity' and 'multiculturalism.' While these terms suggest that something substantive has happened to the cultural make-up of cities, they often have little to say about the actual cultural processes themselves. Rather, 'culture' in these cases is usually subsumed within broader questions about the formation of ethnic and religious communities and how these are implicated in inter- and intra-group relations. Over the past decade, as research on the urban scale of migration has flourished, there has been growing interest in the question of cultural production and consumption in cities. The increasing role of creative industries in urban development, in particular, has led migration scholars to consider the relevance of different cultural practices, such as popular music, in theoretical and policy debates about the incorporation of migrants in cities (Kasinitz & Martiniello, 2019). However, there has yet to be a broader reflection on how a cultural lens might enhance critical knowledge of the relationship between migration and cities.

In order to address the issues raised above and to review the research that has emerged in recent years, this chapter focuses on three areas. First, it considers one particular form of individual expression—creative writing—to reflect on the ways in which migrant and migrant-background writers have represented cities since the Second World War. Second, it examines carnivals and festivals in London, Cologne and Warsaw to illustrate the ways in which migration-themed cultural events are entangled with wider issues of urban change. Third, it develops the idea of migration heritage by analysing the recent trend towards the integration of migration into museum displays and urban tourism.

5.2 Migrant Literatures and the City

Of all forms of human expression, creative writing offers perhaps the most accessible and rewarding means of exploring representations of migration and cities. Writers have long adopted the newcomer's point of view 'as a device for examining the human condition in the metropolis' (White, 1995: 5). Here I focus on texts produced after the Second World War by international migrants or people with a migration background. These first appeared in the 1950s with the publication of novels by writers from British and French colonial territories, such as *The lonely Londoners*, written in creolized English by the Trinidadian Indian Sam Selvon and *Le docker noir* (The black docker), set in Marseille by the Senegalese Ousmane Sembène (both published in 1956). Migrant literatures[4] later developed in central and northern Europe, largely as a legacy of guest worker programmes (Fischer & McGowan, 1995), and from the 1990s in southern Europe with the rise of international migration to that region.[5] Throughout this extensive body of work, the city is almost always present and often has a crucial narrative function. Yet it has received surprisingly little scholarly attention. The numerous monographs and edited studies of migrant literatures over the past three decades have explored motifs such as identity, belonging, exile, memory, racism and gender (Burns, 2013; King et al., 1995; Minnaard, 2009; Sabo, 2018), while the urban dimension, if considered at all, has largely been a secondary concern (but see, for example, Parati, 2017).

Drawing on a selection of texts, my aim here is to highlight key ways in which creative writing can animate and challenge our understanding of migratory processes at the urban scale. Scholars have highlighted how migrant literatures offer powerful insights into a range of issues, from perceptions of home and displacement to family relationships and social attitudes (White, 1995). Literary accounts encourage us to engage with subjective experiences and perspectives that defy the clear-cut group differences or structural constraints through which migration is conventionally studied. They single out issues before they become statistically relevant. They invite us to consider emotional registers—humour, cynicism, irony, anger, indifference and so on—that are usually absent from academic research. At the same time, any analysis must take into account the limits of interpretation, the contexts in which a text is published, the degree of autobiographical input, and the fact that the lives of authors and their characters are never solely shaped by migration.

At the most immediate level, authors provide us with information about the composition of migrant populations in different cities. This can take the form of a passing remark or a detailed sketch, or it may even be the central focus of a text.

[4] The term 'migrant literatures' is used in the plural form to emphasize the heterogeneous nature of the literary works. The term is used for analytical convenience, fully that many writers do not see themselves as 'migrants' or members of a literary subgenre.

[5] Migrant-produced texts are less common in CEE countries where international mobility has been limited, although there is a flourishing 'emigrant literature' produced mainly by young people who have moved to Western Europe since the early 2000s. For a discussion of Polish migrant writers in London, see Plesske and Rostek (2013).

In *Le docker noir*, Ousmane Sembène draws on his own experiences on the docks of Marseille to create the fictional story of Diaw, a Wolof stevedore who is tried and executed after accidentally killing a white woman. The author presents a close-up portrait of a few blocks between the railway station and the port, which after the Second World War had become 'le petit Harlem marseillais' (Sembène, 1956: 77) and '[l']Afrique méridionale de la France' (78); home and stamping ground for members of various West African ethnic groups ('Sarakoles', 'Susus', 'Wolofs', etc.) who, despite their differences, were 'united by a spirit of community and solidarity' (ibid). In a ground-breaking foray into the representational politics of race and place, Sembène juxtaposes this description with mock newspaper clippings announcing the same area as 'the black district [...] inhabited by pimps, thieves and pickpockets' (26) and 'a den of wolves' (28).

Zadie Smith's international bestseller *White teeth* (2000) tells the intertwined stories of three families in Willesden, north-west London: the first mixed white English and black Caribbean, the second Bangladeshi and the third Jewish-Catholic, white, middle class and liberal. Set in the same area as the Grunwick strike (Sect. 3. 1), the novel depicts a 'mature' multicultural metropolis that is at once extraordinary and mundane, where ideas of home and identity are fluid and class and ethnic compositions too complex to be easily reproduced on a map or codified into a single concept. The financialization of London's housing market and the rapid gentrification of Willesden in recent decades means that Smith's is also an account of a city that no longer exists in the same form today.

Alongside paradigmatic migrant cities such as Marseille and London (where a writer like Zadie Smith is now part of the literary canon), creative writing often focuses on secondary cities and migratory routes that are barely registered in historical research. Luigi di Ruscio's *La neve nera di Oslo* (The black snow of Oslo) (2010), for example, traces the author's emigration from the town of Fermo in central Italy in 1957 to work in a nail factory in Oslo, a city with few Italians and even fewer communist worker-poets. In the novel, di Ruscio describes his love-hate relationship with the Scandinavian capital and his grudging assimilation into local life.

The narrative structures of the literary texts help to illuminate the complex and unpredictable nature of urban migration, revealing the links between different stages and the causes and consequences of the choices people make. Through their diachronic perspectives, the texts offer new angles on familiar tropes, such as the shock of arrival in the unfamiliar metropolis, and they complicate assumptions about the significance of the city in people's migration projects. In Sam Selvon's *The lonely Londoners* (1956), a tale about Caribbean male migrants in the Ladbroke Grove area of west London during the 1950s, the main character, Moses Aloetta, is a reluctant mentor to new arrivals from the islands, forever grumbling about his pioneer status and tired of supporting other 'fellers' he has never met. On one level, the book lampoons idealized notions of group solidarity; on another it shows how Moses is trapped by a sense of loyalty and his own quest for respect and monetary return.

Fifty years later, in *An Arab melancholia* (2012), Moroccan writer Abdellah Taïa articulates the interwoven experience of being a migrant and a gay man through his fraught relationship with different cities: his native Salé (with its twin city Rabat);

Cairo, as the cinematic city of childhood dreams; and Paris, where he settles after a period of study in Geneva. These localities are marked by ambivalence: the emotional sustenance, fantasy and excitement of Rabat-Salé and Cairo are offset by the secrecy and violence surrounding homosexuality in North African societies, while the freedoms and career opportunities of Paris and Geneva are counterbalanced by loneliness, dreariness and rejection. In other words, migration to the European city is not simply a one-way ticket to emancipation, love and economic betterment; it can also lead to loss and disillusionment.

Creative writing often alerts us to the different urban worlds, both real and symbolic, that migrants inhabit and move between. In Taïa's work, as in that of many others, the migrant writer acts as a 'comparative urbanist' (Robinson, 2006), placing cities in relation to each other in innovative ways. For example, Ubah Cristina Ali Farah's novel *Il comandante del fiume* (The Commander of the River) (2014) explores the intersections of blackness and place in the life of an Italian-Somali teenager as he moves between Rome and London, set against stories from war-torn Mogadishu. The translocal dimension of urban experience is not a recent development in migrant literature. Indeed, novels written or set in the 1950s and 1960s rarely portray arrival in a European city as something particularly remarkable. In Buchi Emecheta's semi-autobiographical novel *Second-class citizen* (1974), the negative reaction of the protagonist, Adah, to the dreariness of Liverpool and later London on her arrival from West Africa in 1964 is partly prompted by her fond memories of life in Lagos. At the same time, the city can be used to engage with other scales, notably the nation. The characters in *The lonely Londoners* constantly exchange 'old London' for 'Brit'n' and 'the Mother Country', although few move very far from their base in Ladbroke Grove.

Allusions to the importance of translocal networks and the disarticulation of the nation as an interpretive frame demonstrate the capacity of migrant literatures to anticipate theoretical and empirical debates in migration studies. Long before the terms 'superdiversity' and 'hyperdiversity' were coined, literary and autobiographical texts probed the multiple variables that shape the incorporation of migrants in cities. Through the character of Adah, Emecheta highlights the ethnic, gender and class tensions within London's 'Nigerian community' and how these were exacerbated in the new urban context. Thus, Adah is forced to hide her Igbo origins in order to be accepted by her Yoruba landlord, while it is her dual role as sole breadwinner and carer that allows her workshy husband to pursue his studies. If the city is a place where social structures are dismantled, it can also be a place where boundaries are re-established. In Kamala Markandaya's *The nowhere man*, first published in 1972, an ageing Indian-English couple living in a comfortable south London suburb are antagonized by their racist white neighbour, who is unemployed and resentful at losing his place in the social housing queue after his own failed migration project to Australia (Markandaya, 2019 [1972]). When the novel was republished in 2019 after decades of obscurity, it was seen to 'resound eerily' with the resurgent racism following the Brexit referendum, from which global London was not exempt (Garman, 2018).

Migrant literatures produce illuminating gendered perspectives, which can be useful in redressing assumptions about the presence and role of women and men in the city. Emecheta's work, for example, has often been cited by sociologists and historians as a rebuttal to male-centred accounts of early multicultural London (e.g. Moch, 2003). It is important to note that the post-war literary field was dominated by men: this was not simply due to the greater number of male immigrants in the 1950s and 1960s, but was indicative of the unequal gender relations in the publishing industry at the time. A case in point is Beryl Gilroy's *In praise of love and children*, the story of two siblings from British Guiana who arrive in London in the 1950s, and the first British novel to address the issue of interracial marriage in the post-war period (Gilroy, 1996). Written in 1959—three years after the publication of Selvon's *The lonely Londoners* and at a time of growing interest in Caribbean literature—it was not published until 1996, after Gilroy had become well known as London's first black headteacher. Over the past four decades many major writers of migrant origin across Europe have been women, and female perspectives on urban life, both past and present, have become increasingly common. In the fictionalized memoir *Die Brücke vom Goldenen Horn* (The bridge of the Golden Horn) (1998), Emine Sevgi Özdamar recounts her own sexual and political awakening as a teenage Turkish factory worker in Berlin in the late 1960s. The factory, the women's hostel and Berlin itself are all treated with picaresque irreverence that, combined with the protagonist's desire to eventually return to Istanbul to become an actress, offers an alternative portrait of the Gastarbeiter experience to that of the male workers depicted in John Berger and Jean Mohr's *A Seventh Man* (Sect. 3.1).

Whether fictionalized or autobiographical, literary texts can be read as valuable commentaries on the migrant urban condition. Certainly, migrant literatures represent a relatively elite cultural sphere. There is a much larger world of non-professional creative writing that is never published. Especially with the rise of digital communication, the practice of writing—like making music or visual art—offers many people an intimate means of communicating with peers, an identity marker, or even a form of therapy. Nevertheless, with due attention to the contexts of artistic production, literary texts hold immense potential for migration studies, both as sources of inspiration and as sources of data (Longo, 2015). They open up scenarios that are otherwise inaccessible or unthinkable. They can disrupt sociological truisms about cultural diversity, but they can also advance prototypes that are themselves open to dispute. In short, the appeal of creative literature lies in its ability to contribute thick descriptions of the migration-city nexus that 'are often infinitely more subtle and meaningful than studies of migrants which base themselves on cold statistics or on the depersonalised, aggregate responses to questionnaire surveys' (King et al., 1995: x).

5.3 Migration-Related Carnivals and Festivals

The terms 'carnival' and 'festival' are used interchangeably to refer to one-off or recurring cultural events that take place in open public spaces or behind closed doors and are either free to attend or, in the case of many arts festivals, charge admission. Four general points should be first emphasized to help frame discussion.

First, many contemporary European carnivals and festivals can be considered eminently urban phenomena (Finkel & Platt, 2020).[6] This does not simply mean that they are located in a city and rely on its infrastructures and audiences, but that they draw on and contribute to the cultures, identities and lifestyles associated with a city and lead to the resignification of its spaces.

Second, the surge in popularity of urban carnivals and festivals in recent decades has paralleled their increasing appeal to local policy makers and private operators, who have incorporated them into a range of agendas from urban regeneration to community cohesion (ibid). Thus, in addition to their cultural and social attributes, carnivals and festivals are seen to offer economic and political dividends that can be directed towards specific goals.

Third, festivals, and carnivals in particular, have long been important, if temporary, sites of transgression and contestation, where boundaries of authority are tested, conventional roles are reversed and social distance is broken down. This is not to say that carnivals have always been emancipatory spaces: on the contrary, participation in ritualized acts of insubordination was historically seen as a sign of allegiance to the status quo and could involve the public denigration of weaker groups such as women or people with disabilities (Stallybrass & White, 1986). Today, carnivals and festivals continue to provide opportunities for spontaneous or organized dissent, although this dimension is often seen to be undermined by the increasing privatization and securitization of public space (Smith, 2015).

Fourth, festivals and carnivals have always been closely associated with human mobility, from travelling performers and merchants with their unfamiliar wares to contemporary celebrations of multiculturalism (Saeys, 2021; Stoeltje, 1992). As a long-standing core element of festivities, mobility is thus connected with both the presence of people from elsewhere and the cultural paraphernalia produced before and after their arrival.

To understand how these issues play out in relation to migration-related carnivals and festivals, I will briefly discuss three cases: Notting Hill Carnival, London's internationally renowned celebration of Black Caribbean cultures; Cologne Carnival, one of Europe's oldest carnivals, which has incorporated migration themes into its programme; and two small festivals organized by Vietnamese migrants in Warsaw, which point to significant divisions within this particular group.

---

[6] There is, of course, a long tradition of rural and agrarian festivals throughout Europe, although many of these today depend on visitors and services based in cities.

5.3.1 Notting Hill Carnival

Notting Hill Carnival is Europe's largest street event, attracting around two million visitors to the west London district over two days in late August each year. It is famous for its blend of classic and contemporary black music genres, Caribbean-themed floats, flamboyant masquerades and street food such as Jamaican jerk chicken. The origins of the carnival lie in the merger of two separate community-organized events: an indoor Caribbean cultural jamboree, founded in 1959 in response to a spate of racist attacks in Notting Hill the previous year, and an open-air festival, inaugurated in 1966, which aimed to bring together immigrants of different origins and to counteract the stigma surrounding the area (Younge, 2002). With the subsequent introduction of steel pan bands and sound systems, the carnival became a reference point for black Caribbean popular culture, attracting a quarter of a million revellers by the mid-1970s.

Notting Hill Carnival has never been just about entertainment. The year-long preparations, largely coordinated by women, are instrumental in building and maintaining diasporic and community ties (Mistlin, 2021). Borrowing elements of Trinidadian carnival forged from the legacy of slavery, the event itself is a celebration of freedom and fundamentally non-elitist. As its primary constituency shifted from foreign-born immigrants to British-born black people, carnival also became a flashpoint for the over-policing of black communities, and after violent clashes between youths and the police in 1976, the event was seen as having 'political implications beyond the intentions of the organizers' (Pryce, 1985: 38). Until the early 2000s, the carnival regularly received negative coverage in the British mainstream media due to recurring incidents of crime and disorder, and was frequently threatened with closure. Over the past two decades, it has instead been widely promoted as a highlight of London's cultural calendar, a major tourist attraction and a boon to the local economy, as well as a photo opportunity for authority figures to demonstrate their cosmopolitan credentials. This general shift in public opinion reflects the expansion and success of carnival, but also the incorporation of aspects of black music into mainstream popular culture and, crucially, the harnessing of cultural diversity as both intrinsic to London's identity and a tool of urban governance.

Since the 1990s, the site of the carnival has undergone intense gentrification, with many new high-income households renting out their properties or leaving the area during the event. Despite this, organizers have resisted periodic calls to move the carnival to nearby Hyde Park. The fact that it still takes place on the streets of west London, despite the changing demographics of the area, is testament to its enduring symbolic power. Today, as in the past, Notting Hill Carnival is best understood as the two days of the year 'when black people come together to assert their presence culturally, politically, socially […] in an overwhelmingly white milieu' (Pryce, 1985: 36). It continues to provide a collective platform for London's Caribbean diaspora and other local minority groups, as evidenced by the pledges of solidarity for black

Britons caught up in the Windrush scandal[7] and the 72-second silences held during the 2018 and 2019 editions in memory of the 72 mostly migrant and minority residents who died in the fire at Grenfell Tower, the unsafely clad social housing block on the poorer, northern edge of the carnival site. As such, Notting Hill today manages to combine two kinds of carnival: an increasingly establishment-friendly multicultural extravaganza, and a space for community organization and resistance, where the politics of race, migration and urban injustice can be brought to public attention.

5.3.2 Cologne Carnival

Cologne Carnival is a historic, citywide carnival that has played an important role in the formation of local identity in the western German city since the early nineteenth century (DeWaal, 2013). Officially beginning on 11 November and culminating in the six 'crazy days' before Ash Wednesday, the carnival now attracts around one million spectators and is known for its costumed balls, stage shows, excessive alcohol consumption and general debauchery, but also for alternative events, such as the Ghost Parade, which first began as an anti-war protest in 1991 and has become increasingly popular and influential (Gannott, 2004). Since the 1990s, carnival organizers have introduced migration-themed elements into the official programme. With more than a third of the city's one million inhabitants of non-German origin, local authorities have cultivated Cologne's image as a liberal, cosmopolitan city, where even centrist and conservative mayors adopt pro-migrant positions. For example, the design of the official 2008 carnival medal for the Cologne district of Ehrenfeld included the city's new central mosque among other local landmarks.

In her ethnographic study of the Cologne carnival, Monika Salzbrunn suggests that the impetus for incorporating migration issues into the carnival came largely from the creativity of activist groups and migrants themselves (Salzbrunn, 2014). Since 1999, the solidarity network Kein Mensch Ist Illegal (No One is Illegal) has organized an annual parade through the city centre called Keine Jeck es Illejal (No Fool is Illegal), which uses the music and theatrical customs of carnival to parody the positions of local and national politicians. It challenges the official rhetoric of diversity during carnival by highlighting the everyday problems faced by migrants in the absence of adequate cultural and social policies. As Salzbrunn observes, the parade provides undocumented people with a safe space for political and cultural expression that is not available at other times, although the main actors remain German nationals or those with residence permits (ibid: 100). As Keine Jeck es Illejal has become an integral part of carnival and has been publicized in local listings magazines, migration has

[7] The Windrush scandal, which erupted in early 2018, was a consequence of the UK government's 'hostile environment' policy for irregular migrants introduced by Home Secretary Theresa May in 2012. Commonwealth-born individuals who had arrived in the UK as British subjects prior to changes to immigration law in 1973, and who had no documentary evidence of their right to remain in the country other than their landing cards (which were destroyed by the UK government in 2010), were denied access to public services and, in some cases, detained and/or deported.

moved from the margins to the centre of cultural practice, while public authorities, concerned with the city's reputation have been pressured to recognize more radical platforms, even if carnival itself has not led to substantial political change. Funds raised by the event and the sale of associated merchandise have been donated to pro-migrant campaigns, including search and rescue organizations operating in the Mediterranean Sea. The critical capacity of carnival as a cultural public sphere (Giorgi et al., 2011) was reaffirmed in the 2016 edition following the sexual assaults on women perpetrated by men of mainly North African origin in Cologne on New Year's Eve 2015. Against a backdrop of heightened tensions and security measures, feminist groups drew attention to the broader issue of patriarchal violence, while some floats in the official Rose Monday parade parodied far-right organizations seeking to whip up anti-refugee sentiment (Connolly, 2016).

5.3.3 Vietnamese Tết Festivals in Warsaw

The final case concerns two Tết Lunar New Year festivals organized by Vietnamese migrants in Warsaw. Tết is the most important national holiday in Vietnam. For a long time, the only organized event in Warsaw was a formal affair at the country's embassy in the south of the city. In contrast to London and Cologne, the Polish capital—at least until the start of the war in Ukraine—had a small foreign-born population, which officially stood at 25,923 at the end of 2016, accounting for roughly 1.5% of the city's total population (Duszczyk et al., 2019: 135). The number of Vietnamese was close to 3,000, making them the second largest group in the city after Ukrainians (Karpieszuk, 2016). People from Vietnam began migrating to Warsaw during the communist era as state-sponsored students and contract workers, and from the 1990s they played a pioneering role in the development of transnational wholesale markets in the city (Marouda, 2019). Today, this small population is diverse in terms of length of stay, age and legal status. The majority are employed in catering or retail and have links to the Vietnamese state, but there are also a number of people with refugee status who identify with the anti-communist diaspora in the United States and France. The emerging divisions within Warsaw's Vietnamese 'community' are explored by Polish anthropologist Grażyna Szymańska-Matusiewicz (2015) in her ethnographic study of two new Tết festivals, established in 2014.

The first festival was organized by the official Association of Vietnamese in Poland in a primary school on the southern outskirts of Warsaw near the major trading centre of Wolka Kosowka, where many Vietnamese live and work. The event, which consisted of a standard programme of singing, dancing, martial arts and festive food, attracted some 600 people, two-thirds of whom were Vietnamese families and embassy representatives, and the remaining third were representatives of local institutions and Polish schoolchildren with their parents who had been invited to attend. The choice of venue was a deliberate strategy to get away from the stuffy embassy environment and reach out to the children of Vietnamese migrants. According to Szymańska-Matusiewicz, a key aim of the organizers was both to ensure the loyalty of the diaspora to the home country and to create an 'image of an unproblematic and harmonious community, attached to "traditional Vietnamese culture" and national values, but at the same time unthreatening to the Polish majority society' (Szymańska-Matusiewicz, 2015: 62).

The second festival was organized the following day by a group of pro-democracy activists in collaboration with a Polish human rights organization at the headquarters of an important publishing house in the centre of Warsaw. Although it bore some similarities to the first event, such as the festive food and entertainment, the participants were mainly Polish and numbered fewer than 150 people. However, there was a significant presence of second-generation Polish-speaking Vietnamese youth, who were conspicuously absent from the first festival. Here they were attracted by a more appealing line-up, including performances by a Vietnamese-Polish alternative rock band and beatbox artists, and, unlike many of their compatriots at the other event, did not view the pro-democracy activists as unpatriotic.

The two festivals convey different iterations of what it means to be Vietnamese in Poland, from nostalgia for the homeland to the promotion of new urban subcultures. They attract different audiences: the first is predominantly Vietnamese, family-oriented and suburban; the second is mainly Polish and metropolitan, but stimulating enough to attract second-generation interest. As well as being cultural celebrations, the two events are significant for their political leverage. The 'official' Tết seeks to maintain influence over the economically productive Vietnamese population that has moved to the outskirts of Warsaw in recent years, while the 'pro-democracy' Tết is primarily concerned with promoting an alternative image of the community in the minds of the Polish public.

5.3.4 Carnivals and the Migration-City Nexus

As the cases above illustrate, carnivals and festivals reveal the powerful role that the urban environment can play in shaping and transmitting the cultural practices of migrant and migration-background groups. Such events can complicate or reinforce the relationship between ideas of local identity and particular locations in the city. They shed light on the 'uneasy tensions between socio-economic strategies of commercialized neoliberal cities and cultural needs of diverse communities to gather and celebrate' (Finkel & Platt, 2020: 2). But as we have seen in all three cities, carnivals and festivals also provide spaces, however ephemeral, for self-organization, confrontation and resistance.

Exploring the migration-city nexus through carnivals and festivals means more than simply recognizing different forms and expressions of cultural diversity and how they might be mobilized at a policy or community level. As the two Vietnamese festivals in Warsaw demonstrate, certain events are more concerned with defining the characteristics of a particular diasporic group than with contributing to an overall sense of 'diversity.' Conversely, an exclusive focus on 'multicultural' indices risks minimizing the relevance of human mobility, especially if this latter dimension is seen as detracting from the local or commercial character of an event. The power of urban carnivals and festivals lies precisely in their ability to bring to the fore the multiple intersections between migration, settlement, citizenship and modes of representation and reception.

5.4 Migration, Cities and Cultural Heritage

In recent years, cultural heritage has become the focus of a broad field of inquiry that intersects with critical issues facing contemporary post-industrial societies. No longer assumed to be the preserve of connoisseurs and conservationists, today heritage often lies at the centre of debates about belonging and identity, the commodification of culture and the common good. Cities are particularly relevant in this regard, as they are caught between national and local heritage discourses, and between competing versions of the past (Smith, 2006). At the same time, dominant ideas about urban heritage in Europe—whether expert-led or community-based—have tended to rely on a territorial conception of culture, which is seen as firmly *rooted in place*. According to such a scheme, a city's heritage is associated with material vestiges of the past, customs and traditions, which together are seen to contribute to a distinct sense of *local identity*. Indeed, as all cities across Europe are caught up in an increasingly globalized economy, cultural heritage is seen as one of the key things that sets them apart.

Herein lies a problem. As illustrated in Chapter 2, cities have *also* historically been shaped by human mobility, and contemporary processes of globalization have led to an intensification of their translocal connections. This raises a fundamental question: if urban heritage is to be understood as something *rooted in place* and constitutive of *local identity*, how do the histories of migration that characterize each city fit into the equation?

This 'dilemma' makes heritage a particularly interesting analytical lens through which to examine the relationship between migration and cities. Indeed, over the last two decades there have been signs of a shift away from sedentary conceptions of heritage. Here I consider two examples of heritage-making—museums and guided tours—that have recently received attention in the academic literature (Levin, 2016; Ormond & Vietti, 2022; Whitehead et al., 2015). I use the term 'migration heritage' to refer to the ways in which legacies and memories of mobility are mobilized and represented in the present. Migration heritage is not reducible to the artefacts, practices and belief systems that emerge and evolve as a result of migration, but encapsulates how layers of movement have transformed the meanings of localities and how past events and processes resonate with the present. Including migration in the concept of urban heritage also raises the question of the relationship of contemporary migrants to heritage. As newcomers, are migrants expected to learn about and respect the heritage of their host city—perhaps as a sign of their 'integration'—or are they themselves active contributors to heritage from the moment they arrive? This question not only points to competing conceptions of heritage and how these change in different urban contexts, but also leads us to interrogate who and what constitutes the 'cultural diversity' of cities.

5.4.1 Migration in City Museums

One way in which migration has been incorporated into the heritage industries of European cities has been through its representation in museums. The fact that migration did not become a permanent feature of exhibitions until the early 2000s is somewhat remarkable, given its unrelenting significance throughout European history. In large part, this is because museums have traditionally been seen as contributing to *national* culture and identity. This is clearly the case with the various 'national' museums found in major cities, but it is also true of smaller local museums that have situated their displays within a broader national framework. In such contexts, migration is seen as irrelevant or as a de facto threat to the idea of a territorially bounded national community. The growing interest in migration-related themes reflects a broader shift in museum practice in recent decades, with different kinds of museums becoming more inclusive of other communities and their stories, reaching out to new audiences and responding to the cultural challenges of globalization (Levitt, 2015; Whitehead et al., 2015).

For convenience, we can divide the musealization of migration into three basic typologies: stand-alone migration museums; city museums in which migration is a sub-theme of a permanent collection; and temporary exhibitions that explore migration-related themes. The first group includes national institutions such as the National Museum of Immigration History in Paris (opened in 2007) and the Museum of the History of Immigration in Catalonia near Barcelona (opened in 2004), as well as a number of emigration museums located in port cities such as the German Emigration Centre in Bremerhaven (opened in 2005), the Red Star Line Museum in Antwerp (opened in 2013), the Emigration Museum in Gdynia (opened in 2014) and EPIC The Irish Emigration Museum in Dublin (opened in 2016). Many of these new museums have been designed to be major tourist attractions and anchors for urban regeneration projects. Some have been established in new signature buildings, such as the Emigration Centre in Bremerhaven, while others have reused historically significant sites. In Antwerp, for instance, the museum is housed in the former sheds of the Red Star Line shipping company, which between 1873 and 1935 transported two million Europeans to North America. The second group includes city museums such as the Amsterdam Museum and the Museum of London, which have (re)organized their displays to reflect the multiple forms of mobility that have shaped their respective cities over time. Finally, temporary exhibitions on migration have a somewhat longer history and have often paved the way for more sustained engagement with the issue. Many have pioneered community engagement and introduced moments of critical reflection. They have also addressed aspects such as personal and collective memory, which are frequently overlooked in permanent displays.

In their research on the representation of migration in various European museums, Whitehead et al. claim that '*place* is a fundamental epistemological structure and referent within museums', operating 'as a force within identity work; as something inextricably connected with temporality; and as an entity made manifest in material objects' (2015: 7). This 'place', however, is not necessarily urban. Indeed, while most migration museums are located in cities, and some have become urban landmarks, many focus on national histories and diasporas that exclude or obscure any interest in the city itself. This is well illustrated by the Galata Maritime Museum in Genoa, which opened in 2004 as part of the harbour redevelopment and was originally designed to celebrate the city's maritime heritage (Dines, 2017). Since November 2011, the museum's third floor has housed an immersive exhibition on the Italian emigration experience and Genoa's own history as a major port of departure. At the end of the itinerary—after boarding an ocean liner, disembarking at a replica of the Ellis Island arrival hall and viewing panels on the life of Italian emigrants in North and South America—visitors can choose to enter a final small section on contemporary immigration, the only permanent exhibition on the subject in Italy.[8] The first part features a giant timeline of the history of immigration to Italy since the 1970s and an interactive world map with information on contemporary global migration flows. The local context is discreetly introduced in a subsequent room through a visitor-operated console that activates video interviews with international migrants who have settled in Genoa since the 1970s. The main exhibit in this room, and the only physical object in the entire section, is a Tunisian fishing boat donated to the museum by the municipality of Lampedusa, surrounded by looped images of rescue operations in the Mediterranean Sea. The representation of migration in the Galata Museum is thus framed by a series of powerful tropes—national exodus, maritime crossings and humanitarian crisis—that overshadow any engagement with the urban scale. This raises some important questions about the relationship between migration heritage and the museum. First, the presentation of foreign immigration as an appendix to Italian emigration ignores the co-presence of this and other types of migration throughout history. Second, the centrality of the maritime journey obscures the fact that the vast majority of foreign migrants in Genoa—including those interviewed for the exhibition—arrived by air or land, and that the city's port has never been a major point of entry for immigration. Finally, the use of a humanitarian framework obscures the more mundane, local contingencies of immigration and the ways in which migrants have contributed to the reconfiguration of Genoa's cultural landscape. Ironically, just a few metres inland from the Galata Museum is Via del Pré, one of Genoa's long-standing 'migrant streets', which is now home to many businesses run by people from South Asia, North Africa and sub-Saharan Africa. However, the immigration exhibition makes no reference to its existence.

[8] Since the early 2000s, there have been a number of temporary exhibitions on immigration in major cities such as Turin and Rome. In May 2022 the National Museum of Emigration was opened in Genoa across the street from the Galata Museum, although the latter has so far retained its emigration and immigration sections.

The city's presence and significance are similarly marginal in the permanent exhibitions of the above-mentioned emigration museums and nationally oriented institutions (Stevens, 2008; Sutherland, 2014). In contrast, temporary exhibitions have often explicitly explored the urban dimension of migration. For example, 'The Peopling of London', a ground-breaking exhibition held at the Museum of London between 1993 and 1994, challenged the common view of multicultural London as something that developed *after* the Second World War. It illustrated the overlooked history of London's medieval Jewish community, the continuous black presence since the sixteenth century, the establishment of Asian communities from the seventeenth century, and the immigration of Chinese, Italians and Germans in the nineteenth century (Merrimen, 1993; Ross, 2015).

The inclusion of migration in place-based displays faces a number of challenges. Museums must balance the presentation of difficult histories of colonialism, racism and violence with the need to attract a wide range of visitors and meet regulatory and funding requirements. Positive portrayals of migration can provoke political opposition, but can also overlook the realities of disadvantage, prejudice and restrictive legislation that affect some groups of people more than others. Moreover, any attempt to capture the impact of migration on a locality is inevitably selective, which can inadvertently privilege or exclude certain voices. As Whitehead et al. argue, 'multicultural society comes to be represented through samples of diversity, working emblematically rather than thoroughly' (Whitehead et al., 2015: 32).

5.4.2 *Everyday Traces of Migration Heritage and Urban Tourism*

Cultural institutions are not the only places where migration heritage is produced, represented and consumed. Cities are full of reminders of past movements and settlements. Street and neighbourhood names such as Fondaco dei Turchi in Venice, the Dutch Quarter in Colchester and Moldovanka in Odessa bear witness to the origins of historical migrant communities (Chapter 2). Local culinary traditions are often the result of the migration of ingredients, recipes and people. Balti curry, for example, is said to have been invented in Birmingham in the 1970s by Pakistani immigrants. A number of major football clubs—an important marker of local identity for some people—were founded by migrant workers or refugees: Celtic in Glasgow and Hibernian in Edinburgh have Irish connections, while AEK Athens and PAOK Thessaloniki were founded in the 1920s by Greeks from Anatolia. If migration heritage can be understood as something inherited (for instance, through membership of a diasporic community), it is also actively remade, whether through the settlement of urban areas and the re-appropriation of public spaces, or through the establishment of place-specific cultural practices such as dub poetry in Brixton or carnival in Notting Hill.

The extent to which such elements are meaningful depends on people's own experiences and perceptions of migration, the degree of shared knowledge, or the relevance of a past flow at a given moment in time. Certain tangible or intangible signs of past migrations—such as a particular neighbourhood or place of worship—may possess sufficient symbolic significance to be widely acknowledged by a city's population; others may simply be forgotten, or be too inconspicuous or intimate to be meaningful to the wider public. In any case, all cities have a range of migratory traces that can be claimed and reanimated by different groups of people, which in turn can serve to counterbalance static notions of place and heritage, but also to rewrite assumptions about the times and spaces of migration and diversity.

In line with a heightened sensitivity and interest in the complex diversity of European cities, recent years have seen a proliferation of institutional and grassroots initiatives aimed at activating migration heritage, many of which rely on the internet and social media for their presentation and dissemination. An increasingly popular format is the guided tour. A notable example is the Migrantour network, which coordinates tours of neighbourhoods associated with migration in cities in Italy, France, Spain, Portugal, Belgium, the Netherlands and Slovenia.[9] Launched in Turin in 2009 and born out of a collaboration between anthropologists and operators in the field of 'responsible tourism', these tours have aimed to illustrate and valorize the contribution of different generations of migrants to the transformation of urban areas, which were seen to be generally absent from public discussions about contemporary cites (Ormond & Vietti, 2022; Vietti, 2018). Local people with migrant backgrounds are trained as guides and involved in designing the itineraries in each participating city. Each tour includes visits to shops, eateries, community spaces and places of historical significance (some of which may no longer have a visible link to migration), biographical information about the guides and encounters with other migrants and locals. The project attracted an estimated 30,000 participants in its first decade and received widespread media attention, including some disapproval from right-wing politicians (Vietti, 2018).

The main tour in Naples since 2016 is organized around Piazza Garibaldi, the same public space discussed in the section on open public space in Sect. 4.4.1. The tour takes place up to three times a week, usually on request. The guides are Senegalese, Sri Lankan, Ecuadorian and Ukrainian members of a local association who work as cultural mediators for the city council. The tour lasts about two hours and stops at a number of local shops and businesses, including a Tunisian halal butcher-cum-teahouse, a Pakistani-Senegalese fast-food restaurant, Naples' oldest international supermarket, an Egyptian-run store specializing in African and Asian jewellery, and a Neapolitan pastry shop that has been producing North African sweets for local customers since the early 2000s. The tour also visits a Senegalese mosque and less obvious places, such as an unassuming alleyway known to many migrants as the 'Somali street' because it was once the meeting point for Somalis passing through Naples.

[9] For a full list of participating cities, see: http://www.mygrantour.org/category/citta-migrantour/

The four Migrantours around Piazza Garibaldi that I attended between 2016 and 2024 were characterized by a matter-of-fact approach to migration and diversity that avoided an overly celebratory tone but did not shy away from addressing sensitive issues, such as the rise in Islamophobia experienced by worshippers at the mosque. The tours presented the railway station district in a different light to the negative portrayals of the same area in the local media, which have associated migrants with insecurity and urban blight. At the same time, the Migrantour experience raises two pressing questions. First, what and whose migration heritage is being produced during these tours? And, second, who is their target audience? With respect to the first question, Migrantour is coordinated by white Italians, although migrants play a decisive role in shaping the style and content of the tour. In Naples, this meant emphasizing the interactions between different groups, including 'native' Neapolitans, and refraining from presenting people, places and activities as representative of a particular 'ethnic community' or local multiculture. Nevertheless, the focus was mainly on sub-Saharan and North African male migrants, who have a highly visible presence in the station area, but who actually represent a small minority of the city's total migrant population. Most migrants in Naples, especially the many eastern European women employed in the domestic and service sectors, have a more intermittent relationship with urban public space and were therefore not included in Migrantour. In response to the second question, Migrantour was publicized to international tourists through online advertising, including on short-term rental platforms, but in the case of Naples, as elsewhere (Vietti, 2018), the tours were mainly attended by local middle-class residents and school and university students on field trips. Audiences in Naples were, in a sense, offered a staged, if not particularly polished, version of migration heritage in the station area. The 'backstage' of the tour—how the various 'characters' encountered along the way experience migration in Piazza Garibaldi and perceive their own contribution to its transformation over time—is likely to be messy and far more complex.

The project's publicity is frank about the ambivalence of migrant tourism, stating in its brochure that 'the links between migration, tourism and cities are a complex phenomenon with rich potential but also risks that must be confronted with knowledge and awareness of their symbolic and political meaning' (Migrantour, 2015: 2010). Indeed, according to its coordinators, Migrantour is more than just a form of alternative tourism or a source of income for its guides: it is a mode of 'civic learning' and a means of building more inclusive heritage communities (Ormond & Vietti, 2022). There has also been a lively academic debate around Migrantour, with criticisms of the project's lack of attention to racism, urban inequalities or the negative impacts of overtourism and gentrification (Mellino & Vietti, 2019; Pozzi & Ceschi, 2019; Vietti, 2018). In response, organizers have taken a more reflexive approach to the limitations of representing cultural diversity, and have encouraged guides to engage with situations of conflict and their own intersectional differences on their own terms.

Despite being at the centre of critical attention, Migrantour is not solely responsible for the production of migrant heritage in Naples or other cities involved in the project, but is one of a plethora of practices that are differently situated in relation to migration histories and take multiple forms, including variations on the tour format.[10] Heritage-making is not an accumulative process in which new layers are gradually added to a substrata of meaning: it is an ongoing, contested and mutable project. In the coming years, and in parallel with the growing interest in rethinking urban histories in relation to colonialism, slavery and racism, migration heritage is likely to become an important focus of urban research, especially in those localities where migration-driven diversity remains marginal to public notions of local identity.

5.5 Conclusion

Cultural processes have been surprisingly neglected in research on migration and cities, despite the almost ubiquitous references to 'multiculturalism' and 'cultural diversity' in the literature. Linton Kwesi Johnson's dub poetry and Phaim Bhuiyan's *Bangla* are as much an outcome of the migration-city nexus as any economic or settlement process. While these latter processes may inform our thinking about how migration shapes urban development or the transformation of the built environment, different forms of cultural production generate nuanced and unpredictable insights into sense-making processes that both contribute to and unsettle our understanding of the links between migration and cities. Johnson's simmering rage at everyday police harassment and *Bangla*'s light-hearted love story set against the backdrop of creeping gentrification produce, each in their own way, commentaries on the city and what it means to belong to it. They remind us that migrants are not simply carriers of 'culture' that replenish the diversity of cities, but active agents in the production, definition and consumption of constantly evolving, plural and often discordant urban cultures.

The three selected areas discussed in this chapter—creative writing, festivals and cultural heritage—signal different entry points for engaging with creative and sense-making processes. In contrast to the other chapters, this has meant paying closer attention to individual examples in order to draw out issues for general consideration. Migrant and migrant-background novelists provide us with an immensely valuable and variable source of 'insider' information, the carnival draws attention to the contradictory intersections between the representation and reception of migration-driven diversities, while different formulations of migration heritage point to the significant links between urban history and local identity, and how these are mobilized and contested in productive activities such as tourism. As Marco Martiniello (2015) notes, the increased interest in the 'arts' among migration scholars has coincided with the growing importance of cultural processes within the broader economic

---

[10] An example of a very different guided tour is 'Harnet Streets—Eritrean counter maps in Rome', launched in 2021. Instead of visiting migrant spaces, the tour and online podcasts are structured around interviews with Eritrean migrants as they walk through the Quartiere Africano, an upper-income neighbourhood in northern Rome whose streets are named after nations and cities that were once Italian colonies (see https://tezeta.it). These interviews focus on people's different relationships with the city of Rome, memories of home and the (post)colonial links between Italy and Eritrea.

and policy dynamics of cities. However, while migrant cultural practices may be use to promote economic incorporation or overcome group divisions, they do not necessarily change things in the long term. Migrant writers negotiate the vagaries of the publishing industry and a dwindling, often elite, readership. Street festivals are subject to the pressures of political expediency and the commodification of the carnivalesque. The survival of a pioneering museum may depend on fickle and politically allocated sources of funding, while in some cities migrant tourism is implicated in the very same processes that lead to residential displacement.

Yet the questions at stake in the migration-city nexus are not reducible to economic productivity or settlement: cultural processes are key ways in which we can begin to confront questions of agency and subjectivity in urban processes, better understand the intricacies of belonging and how the meanings of cities are altered for good. As such, novels, festivals and heritage, among many other things, need to be taken seriously.

References

Ali Farah, U. C. (2014). *Il comandante del fiume* (The Commander of the River). 66th and 2nd.

Appadurai, A. (1996). *Modernity at large: Cultural dimensions of globalization.* Minnesota University Press.

Bhabha, H. (1994). *The location of culture.* Routledge.

Brah, A. (1996). *Cartographies of diaspora: Contesting identities.* Routledge.

Burns, J. (2013). *Migrant imaginaries. Figures in Italian migration literature.* Peter Lang.

Çağlar, A. (1997). Hyphenated identities and the limits of "culture." In P. Werbner & T. Modood (Eds.), *The politics of multiculturalism in the New Europe: Racism, identity, community* (pp. 169–186). Zed Books.

Caroselli, A., Meo, E., & Mellino, M. (2019, July 23). Bangla: Antirazzismo e nuove italianità ai tempi del branding (Bangla: Anti-racism and new Italianisms in times of branding). *Dinamo Press.* https://www.dinamopress.it/news/bangla-antirazzismo-nuove-italianita-ai-tempi-del-branding/#_ftn1

Connolly, K. (2016, February 8). Cologne carnival: "After New Year, there's even more reason to let our hair down." *The Guardian.* https://www.theguardian.com/world/2016/feb/08/cologne-carnival-after-new-year-theres-even-more-reason-to-let-our-hair-down

DeWaal, J. (2013). The reinvention of tradition: Form, meaning, and local identity in modern Cologne carnival. *Central European History, 46*(3), 495–532.

Dines, N. (2017). *Displaced by the humanitarian sea? The representation of contemporary migration in an Italian maritime-emigration museum.* http://epapers.bham.ac.uk/3178/1/Conference_Proceedings_-_Heritages_of_Migration%2C_Moving_Stories%2C_Objects_and_Home.pdf

Di Ruscio, L. (2010). *La neve nera di Oslo* (The black snow of Oslo). Ediesse.

Duszczyk, M., Pszczółkowska, D., & Wach, D. (2019). Warsaw: A new immigration city in search of its integration policy towards newcomers. In T. Caponio, P. Scholten, & R. Zapata-Barrero (Eds.), *The Routledge handbook of the governance of migration and diversity in cities* (pp. 130–142). Routledge.

Emecheta, B. (1974). *Second-class citizen.* Allison and Busby.

Finkel, R., & Platt, L. (2020). Cultural festivals and the city. *Geography Compass, 14*(9), 1–12.

Fischer, S., & McGowan, M. (1995). From *Pappkoffer* to pluralism: Migrant writing in the German Federal Republic. In R. King, P. Connell, & P. White (Eds.), *Writing across worlds: Literature and migration* (pp. 39–56). Routledge.

Gannott, S. (2004, February 20). Karnevalsdialektik: Saufen für den guten Zweck (Carnival dialectics: boozing for a good cause). *Die Tageszeitung*. https://taz.de/Karnevalsdialektik-Saufen-fuer-den-guten-Zweck/!786747/

Garman, E. (2018, November 6). Feminize your canon: Kamala Markandaya, *The Paris Review*. https://www.theparisreview.org/blog/2018/11/06/feminize-your-canon-kamala-markandaya/

Gilroy, B. (1996). *In praise of love and children*. Peepal Tree Press.

Gilroy, P. (1993). *The Black Atlantic: Modernity and double consciousness*. Verso.

Giorgi, L., Sassatelli, M., & Delanty, G. (Eds.). (2011). *Festivals and the Cultural Public Sphere*. Routledge.

Hall, S. (2021 [2006]). Cosmopolitan promises, multicultural realities. In P. Gilroy & R. Wilson Gilmore (Eds.), *Stuart Hall: Selected writings on race and difference* (pp. 386–408). Duke University Press.

Jaggi, M. (2002, May 4). Poet on the front line. *The Guardian*. https://www.theguardian.com/books/2002/may/04/poetry.books

James, M. (2015). *Urban multiculture: Youth, politics and cultural transformation in a global city*. Palgrave Macmillan.

Johnson, L. K. (1991). *Tings an times: Selected poems*. Bloodaxe Books.

Karpieszuk, W. (2016, March 17). Warszawa lubiana przez cudzoziemców. Ilu ich mieszka w stolicy? (Warsaw is liked by foreigners: how many live in the capital?) *Gazeta Wyborcza*. https://warszawa.wyborcza.pl/warszawa/7,34862,19778457,warszawa-lubiana-przez-cudzoziemcow-ilu-ich-mieszka-w-stolicy.html

Kasinitz, P., & Martiniello, M. (2019). Music, migration and the city. *Ethnic and Racial Studies, 42*(6), 857–864.

King, R., Connell, P., & White, P. (Eds.). (1995). *Writing across worlds: Literature and migration*. Routledge.

Lees, L. (2016). Gentrification, race and ethnicity: Towards a global research agenda? *City & Community, 15*(3), 208–214.

Levin, A. K. (2016). *Global mobilities: Refugees, exiles, and immigrants in museums and archives*. Routledge.

Levitt, P. (2015). *Artifacts and allegiances: How museums put the nation and the world on display*. University of California Press.

Longo, M. (2015). *Fiction and social reality: Literature and narrative as sociological resources*. Ashgate.

Markandaya, K. (2019 [1972]). *The nowhere man*. Small Axes.

Marouda, M. (2019). A tale of three marketplaces: Chinese commodities, European fairs, Vietnamese entrepreneurs. *Global Networks, 20*(4), 725–745.

Martiniello, M. (2015). Immigrants, ethnicized minorities and the arts: A relatively neglected research area. *Ethnic and Racial Studies, 38*(8), 1229–1235.

Mellino, M., & Vietti, F. (2019). L'antropologia applicata tra "tecniche di mercato" e "pratiche politiche": Riflessioni sui migranti, Migrantour e noi (Applied anthropology between "market techniques" and "political practices": Reflections on migrants, Migrantour and us). *Antropologia Pubblica, 5*(1), 123–132.

Merriman, N. (1993). *The peopling of London: Fifteen thousand years of settlement from overseas*. Museum of London.

Migrantour. (2015). *Migrantour, the world within cities*. New Press.

Minnaard, L. (2009). *New Germans, New Dutch: Literary interventions*. Amsterdam University Press.

Mistlin, S. (2021, August 18). We brought colour to this country: The matriarchs of Notting Hill Carnival. *The Guardian*. https://www.theguardian.com/culture/2021/aug/18/we-brought-colour-to-this-country-the-matriarchs-of-notting-hill-carnival

Moch, L. P. (2003). *Moving Europeans: Migration in Western Europe since 1650*. Indiana University Press.

Ormond, M., & Vietti, F. (2022). Beyond multicultural "tolerance": Guided tours and guidebooks as transformative tools for civic learning. *Journal of Sustainable Tourism, 30*(2–3), 533–549.
Özdamar, E. S. (1998). *Die Brücke vom Goldenen Horn* (Bridge of the Golden Horn). Kiepenheuer und Witsch.
Parati, G. (2017). *Migrant writers and urban space in Italy: Proximities and affect in literature and film*. Palgrave Macmillan.
Plesske, N., & Rostek, J. (2013). Rubble or resurrection: Contextualising London literature by Polish migrants to the UK. *The Literary London Journal, 10*(2). http://www.literarylondon.org/london-journal/autumn2013/plesskeandrostek.html
Pozzi, G., & Ceschi, S. (2019). L'antropologia applicata tra 'tecniche di mercato' e 'pratiche politiche'. Riflessioni sui migranti, Migrantour e noi (Applied anthropology between 'market techniques' and 'political practices'. Reflections on migrants, Migrantour and us). *Antropologia Pubblica, 5*(2), 127–144.
Pryce, E. A. (1985). The Notting Hill Gate Carnival: Black politics, resistance, and leadership 1976–1978. *Caribbean Quarterly, 31*(2), 35–52.
Robinson, J. (2006). *Ordinary cities: Between modernity and development*. Routledge.
Ross, C. (2015). From migration to diversity and beyond: The Museum of London approach. In C. Whitehead, K. Lloyd, S. Eckersley, & R. Mason (Eds.), *Museums, migration and identity in Europe: Peoples, places and identities* (pp. 61–80). Routledge.
Sabo, O. (2018). *The migrant canon in twenty-first-century France*. University of Nebraska Press.
Saeys, A. (2021). Urban multicultural festivals: Spectacles of diversity or emancipatory events? *Journal of Intercultural Studies, 42*(5), 627–642.
Salzbrunn, M. (2014). How diverse is Cologne carnival? How migrants appropriate popular art spaces. *Identities: Global Studies in Culture and Power, 21*(1), 92–106.
Selvon, S. (1979 [1956]). *The lonely Londoners*. Longman.
Sembene, O. (1956). *Le docker noir* (The black docker). Debresse.
Smith, A. (2015). *Events in the city: Using public spaces as event venues*. Routledge.
Smith, L. (2006). *Uses of heritage*. Routledge.
Smith, Z. (2000). *White teeth*. Penguin.
Stallybrass, P., & White, A. (1986). *The politics and poetics of transgression*. Cornell University Press.
Stevens, M. (2008). Immigrants into citizens: Ideology and nation-building in the *Cité nationale de l'histoire de l'immigration*. *Museological Review, 13*, 57–73.
Stoeltje, B. J. (1992). Festival. In R. Bauman (Ed.), *Folklore, cultural performances, and popular Entertainments* (pp. 261–271). Oxford University Press.
Sutherland, C. (2014). Leaving and longing: Migration museums as nation-building sites. *Museum & Society, 12*(2), 118–131.
Szymańska-Matusiewicz, G. (2015). The two Tết festivals: Transnational connections and internal diversity of the Vietnamese community in Poland. *Central and Eastern European Migration Review, 4*(1), 53–65.
Taïa, A. (2012). *An Arab melancholia*. Semiotext(e).
Vietti, F. (2018). Migrantour – Intercultural Urban Routes: Un progetto di antropologia applicate tra migrazioni, turismo e patrimonio culturale. *Antropologia Pubblica, 4*(2), 125–140.
White, P. (1995). Geography, literature and migration. In R. King, J. Connell, & P. White (Eds.), *Writing across worlds. Literature and migration* (pp. 1–19). Routledge.
Whitehead, C., Lloyd, K., Eckersley, S., & Mason, R. (Eds.). (2015). *Museums, migration and identity in Europe: Peoples, places and identities*. Routledge.
Williams, R. (1983). *Keywords: A vocabulary of culture and society*. Oxford University Press.
Younge, G. (2002, August 17). The politics of partying. *The Guardian*. https://www.theguardian.com/culture/2002/aug/17/nottinghillcarnival2002.nottinghillcarnival

Chapter 6
The Urban Dimension of Migration Policies and Politics

6.1 Introduction

Launched in Toronto in 2008, the Cities of Migration project was designed as a forum for local governments, community leaders, policymakers and businesses to share innovative practices from cities around the world. The initiative was a pioneering example of the transnational city networks that had proliferated in the 2010s in response to growing interest in the urban dimension of migration governance. For its 2014 Berlin conference, Cities of Migration invited the Chair of the World Economic Forum's Global Future Council on Migration, Khaled Koser, to present his view on the significance of cities (Koser, 2014). In his speech, Koser argued that cities 'have a potential to make a difference: a difference to the discourse on migration, a difference to migration policy and perhaps most importantly a difference to the lives of migrants themselves.' However, urban reality was complex:

> Cities represent the best and the worst of migration: migrant entrepreneurs and migrant exploitation; innovative migrants and irregular migrants; diversity and discrimination; hope and hatred [...] But cities in general, over and over again, have been able to move the needle from the negative to the positive, and to focus on opportunities and not just challenges.

Relatively unencumbered by the 'obstacle of sovereignty' facing national governments, cities were in a better position to address migration policy, to learn from each other's best practices and to promote an 'objective' debate. In a rousing finale, Koser declared:

> While states are building walls, cities are building bridges, while states are launching patrol boats, cities are launching ideas, and while states are becoming increasingly unilateral, cities are becoming increasingly transnational.

With his emphasis on practical solutions and the need to 'move beyond a talking shop', Koser's polished oratory captured many of the elements that have made the city such an alluring space for policymakers. The premise is that the city is not only

N. Dines, *Migration and European Cities*, IMISCOE Research Series,
https://doi.org/10.1007/978-3-032-15379-1_6

a key arena for policy discussions, but also a place where these discussions can have a tangible impact.

Syntagma Square has long been an emblematic site of urban protest in Athens. Located in front of the Greek Parliament, it was particularly central during the massive popular uprising against national austerity measures in 2011 (Kaika & Karaliotas, 2017). In late 2014, the same space was turned into a month-long camp by hundreds of asylum seekers, some of whom went on hunger strike in protest at their lack of access to healthcare and travel documents. Three years later, the square was occupied again, this time by a smaller group of Syrian hunger strikers demanding to be reunited with their families in northern Europe. A space that had been the focus of global media attention at the height of the Greek debt crisis was now taken over by migrants, along with their tents and handwritten signs, to voice their grievances and call for action from the authorities. As with the anti-austerity movement, these mobilizations did not begin and end in Syntagma Square. Behind and beyond the decision to protest in public space lay a series of organizational processes and relationships. The strategic choice of the square sought to give visibility to the demands and (im)mobilities of the hunger strikers who mostly lived in peripheral camps and had fleeting encounters with the city centre.

At the same time, the two occupations reflected different sets of circumstances. The 2014 hunger strike, although larger in scale, was somewhat isolated and received limited support from the various leftist movements in Athens. In fact, a separate contemporaneous hunger strike by an anarchist prisoner received far more local attention, while pro-migrant solidarity at the time was largely articulated through resistance to the anti-immigrant raids in central Athens coordinated by the neo-Nazi political party Golden Dawn (Margaronis, 2014). The smaller hunger strike of 2017 must instead be seen in the context of the refugee solidarity movement that rapidly developed in Athens following the sharp increase in sea crossings between Turkey and Greece in 2015. This brought together local Greek activists and newly arrived migrants to create a series of alternative infrastructures and direct actions, including a network of refugee squats, some of which actively supported the protest in Syntagma Square (Kotronaki, 2018).

The smooth delivery of Khalid Koser in Berlin and the extreme actions of the hunger strikers in Athens inhabit very different urban worlds. Koser evokes the city as a policy-friendly environment, while the asylum seekers transform it into an agonistic and disruptive space. The former envisions this environment as conducive to more effective management of international migration. The latter, on the other hand, claim their right to the city, not only as a common space of struggle and resources, but also as a place of transit. As we will see, migration governance at the urban scale is far messier than someone like Koser would have us believe, while 'grassroots' migrant mobilizations are often as much about collaboration and compromise as they are about self-organization or insubordination. Nevertheless, both examples encourage us to think about the connections between migration policy and migrant politics in urban contexts and how these are entangled with other (national and global) frames of reference. Moreover, in both cases the city acts as a kind of incubator: of 'good governance' and 'objective debate' on the one hand, and of political action and coalitions on the other.

This chapter examines the migration-city nexus through the twin lenses of policy and politics. It begins by outlining recent theoretical debates on urban citizenship, and demonstrating how this broad issue intersects with both migration policy and migrant politics. Turning to the field of local policymaking, the chapter considers the historical shifts in approaches to migration governance, discusses the different ways in which 'integration' and 'diversity' are managed at the urban scale, reviews recent lines of inquiry, and critically addresses provisions for undocumented migrants such as 'sanctuary cities.' The chapter proceeds to analyze migrants' political mobilizations in urban contexts and concludes by highlighting the role of cities in anti-migration politics.

In order to provide a critical framework for thinking across the different dimensions addressed in this chapter, it is important to introduce some caveats to the discussion. These are paramount because much of the recent scholarly interest in migration and cities has been filtered through reflections on 'local' politics and, in particular, 'local' policy.

First, we must be wary of anthropomorphizing the city into a subject with its own coherent agency. Khalid Koser, for example, speaks effortlessly of cities 'building', 'launching', 'promoting' and so on. But 'the city' does not *design* and *implement* policies nor does it *do* politics: these are the outcomes of constellations of actors and processes operating in and through urban contexts. No less problematic is the tendency in policy discussions to use the name of a city as a proxy for municipal government, often personified in the figure of the mayor. Such a move runs the risk of overlooking, among other things, the complex power geometries inside and outside administrations that influence the design and implementation of migration-related measures.

This leads directly to the second crucial point about how the 'city' is conceived politically. Politics and policy are not confined within municipal boundaries. As geographers have long insisted, analysis needs to move beyond 'territorialist' readings and attend to the relational and multiscalar configurations of governance (Jonas et al., 2018). This does not mean abandoning the idea of the city as a particular locality, insofar as 'a political moment is always placed, localized and invariably operative in public space' (Dikeç & Swyngedouw, 2017: 11). The tension between scale and localization calls for a more careful use of the terms 'local' and 'urban'. The two are not necessarily interchangeable. For example, 'local policy' typically implies a policy that emanates from a particular (lower) level of government, which may be located in a rural context and thus have little to do with cities. 'Urban policy', on the other hand, refers to a specific policy area (the urban). Indeed, it is possible to distinguish between 'local urban policy' and 'national urban policy.'

Third, any analysis of migration-related policies and political mobilizations in cities cannot be read in isolation from broader contexts of urban and political change. In particular, new forms of entrepreneurial governance coupled with the marketization of public services in European cities since the 1980s (Harvey, 1989), have led to undemocratic and socially exclusive neoliberal development agendas that have had detrimental social effects on large swathes of the urban population and have severely constrained the scope for action in other policy areas. Similarly, we should not overlook the existence of attempts to formulate alternative paths of urban development 'characterized as less growth-oriented and more redistributive and social democratic in intent (if not necessarily in effect)' (Jonas et al., 2018: 5), such as the 'new municipalist' approach of the former Ada Colau administration in Barcelona (Russell, 2019), and how these might reshape the politics of migration.

Fourth, we need to query popular assertions about the propensity and capacity of urban governance arrangements to accommodate migration and diversity. This view reflects what critics have termed the 'local trap': the habit of some scholars and policymakers to assume something inherently 'good' about the local scale, especially with regards to democracy (Purcell, 2006). The idea, for example, that cities represent an antidote to xenophobic populist impulses at the national level (Muggah & Glenny, 2017) tends to disregard the ways in which the urban arena is implicated in the racialization of certain groups and the production of anti-migration politics, such as the Golden Dawn pogroms in central Athens. Moreover, the common claim that the city is best placed to provide practical solutions to migration-related issues neglects the multiple power dynamics that determine what can and cannot be done at the urban scale. An illustrative example is the #LondonIsOpen campaign launched by Mayor Sadiq Khan shortly after the 2016 referendum on the UK's membership of the European Union, which sought to reassure London's one million foreign nationals that they would always be welcome, but inadvertently reaffirmed the ineffectiveness of the global metropolis in changing the course of Brexit and its implications for movement to and from the city.

Finally, as we move forward, it is important to keep in mind two fundamental questions, namely *who* are the targets of 'migration policy' and *who* are the subjects of 'migrant politics'? The term 'migrant' is used in this chapter, as in the rest of the book, as a convenient shorthand to encapsulate a myriad of subject positions and migratory contexts. It is important to remember, however, that the policies and politics reviewed here do not apply to everyone who has moved to cities. Local 'integration' policies, for instance, do not apply to all foreign-born newcomers, but only to particular groups of (mainly non-white) migrants who arrive under particular circumstances, while many of their provisions overlap with the needs and interests of other marginalized sections of a settled population. Contemplating the *who* of migration policy and migrant politics thus prompts a deeper reflection on the conditions under which migration takes place and the asymmetrical power relations that structure urban life.

6.2 Theoretical Debates on Urban Citizenship

Over the last twenty years, the city has become the focus of theoretical debates on political issues previously associated primarily with the nation-state. Some theorists have invited us to 'see like a city', in other words to move away from the illusion of the sovereign that comes with the perspective of the state, and to engage ontologically with a multiplicity of authorities and forms of self-government competing for influence (Magnusson, 2011). Others have examined the city as a political entity that is at once relatively distinct from the nation and deeply imbricated with state power (Harvey, 2012). In relation to migration, the city can therefore represent an arena for interrogating state sovereignty over mobility and citizenship, but also the everyday space in which state border controls are enacted.

In an important early contribution, Reinar Bauböck reflected on the increasing global connectivity of capital, people and cities to call for the emancipation of urban citizenship from the constraints of nation-centred conceptions of political community in order to redefine borders, membership and rights and to strengthen self-government (Bauböck, 2003). He argues for the establishment of a formal status of local citizenship based on residence rather than nationality. By guaranteeing local voting rights to non-nationals and challenging the national monopoly on immigration, urban citizenship has the capacity to become a 'homebase for cosmopolitan democracy' (ibid.: 157) that could mitigate class and racial divisions in cities. Bauböck is by no means an 'urban triumphalist': he does not see local citizenship as replacing the nation-state as the new basis of political life, but rather as complementing national citizenship in a multilevel political architecture necessary to tackle multilevel problems (Bauböck & Orgad, 2020: 1–5).

Other theorists have examined the city as a political institution specifically in relation to migration. The political philosopher Avner De Shalit (2018) starts from a number of premises: immigration (his choice of terminology) is fundamentally urban; immigrants are already 'city-zens' before they become citizens of a nation-state; the 'ethos' of a city is shaped by the realities of coexistence; and local political actors are therefore best placed to address the challenges of 'immigrant integration.' He goes on to consider whether cities should have autonomous immigration policies and whether immigrants should be granted local political rights. In short, De Shalit argues for selective admission policies that would deny residency only to convicted criminals and racists, which he justifies on communitarian grounds as a means of safeguarding the common interests of city-zens and facilitating successful integration. Meanwhile, local political rights should be granted to all residents who are committed to making the city their home, since having a say in its direction and future is also an essential condition for strengthening a sense of place and belonging.

In contrast to optimistic accounts of the political potential of cities, critics have raised doubts about the ability of local polities to decouple themselves from the grip of the nation-state. Indeed, many cities do not have access to the state-controlled revenues necessary to meet the welfare and infrastructure needs of their residents (Kohn, 2020). Researchers have also highlighted how urban citizenship is implicated in the production of particular forms of discrimination. Enrico Gargiulo has analyzed how residency acts as a de facto instrument of exclusion, dividing the population into residents with local membership rights and unregistered or deregistered denizens who face administrative barriers in accessing public services (Gargiulo, 2017). Such a situation has become particularly pronounced in Italy over the last fifteen years, where local authorities have deliberately tightened the legal requirements for obtaining residency (ibid.), and where recently-arrived asylum seekers are required to register their residency in reception centres, thus excluding them from membership rights in the municipalities where they eventually move to live and work (Dines, 2023). Scholars have also seen urban citizenship as a mode of governmentality that shapes the conduct of new residents by distinguishing between 'good' citizens (as civic-minded and hardworking) and 'bad' citizens (as uncooperative and unproductive), rewarding the former and penalizing the latter (Hoekstra, 2015).

In addition to interrogating the mechanisms of local governance, critical theorists have also explored how migrants enact urban citizenship in ways that transcend the 'linear, formal and legal language of status [and] rights' (Isin & Nielsen, 2008: 11). In these approaches, the city represents the arena in which political claims and contestations are articulated, regardless of one's status or inclusion in institutional channels of participation, and regardless of whether one seeks full belonging in the city. By engaging in 'acts of citizenship', subjects—such as the hunger strikers in Syntagma Square—'constitute themselves as citizens or, better still, those to whom the right to have rights is due' (ibid.: 2), even when these same subjects—again, like the strikers—'aspire to better futures elsewhere' (Thomaz, 2022: 214). Other theorists have developed similar inquiries by drawing on Henri Lefebvre's notion of the 'right to the city' (see, for example, Bauder, 2016). The right to the city is understood as 'far more than a right of individual or group access to the resources that the city embodies: it is a right to change and reinvent the city' (Harvey, 2012: 4). Although Lefebvre did not directly address migration, his conceptualization of the right to the city is seen as highly suggestive in that it is linked to use value rather than exchange value, and thus to one's presence in the city rather than to property ownership or formal citizenship. Others have put forward more expansive readings of the 'right to the city' framework in order to spotlight questions such as racialized and patriarchal power relations (Garbin & Millington, 2018; Kofman, 2024), which are often overlooked in theoretical discussions of urban citizenship, including those that focus on the (abstract) figure of the migrant.

6.3 Migration Policymaking and Cities

In parallel with theoretical work on urban citizenship, migration scholars have developed a keen interest in local policymaking. If *immigration policy*—the legal framework that regulates entry and exit from a territory—continues to be determined primarily at the national and supranational levels of government, *migration policies*—the courses of action that address the incorporation of migrants within a given jurisdiction—tend instead to take place in the localities where migrants end up living. In most cases, this means cities.

Migration policymaking at the local level is a multifaceted and often elusive process. It is influenced by a host of factors: from regulations and conditions imposed by central government to local administrative capacity; from the needs of new arrivals to pressures from existing residents. What counts as 'migration policy' consists in a mixture of actions and discourses: from tailor-made programmes to ad hoc measures; from provisions specifically targeted at foreign citizens to universal interventions that also benefit migrants. The design of policy is never solely decided by local governments, but may also involve other actors such as policy advisors, social services, businesses, the local media and a host of civil society organizations from trade unions and charities to social movements and migrant associations. Moreover, what actually happens in practice is as much the outcome of informal actions, behind-the-scenes conflicts, mediation and compromises as it is the result of official policies and formal democratic procedures (Dimitriadis et al., 2021).

Various conceptual frameworks have been used to try to make sense of the complexity of local migration policymaking. Many have stressed the importance of examining the interconnections between 'top-down' and 'bottom-up' processes; between the local implementation of national policies on the one hand, and the influence of local policies on national approaches on the other (Caponio & Borkert, 2010; Filomeno, 2017). A common concept among political scientists is that of 'multilevel governance', which refers both to the ways in which local governance relates to and is shaped by other national and supranational levels of government, and to the horizontal interactions between different governmental and non-governmental actors (Caponio & Jones-Correa, 2018; Scholten & Penninx, 2016). Urban geographers, on the other hand, tend to favour a 'multiscalar' perspective in order to draw attention to the links between migration governance and urban development strategies, as well as to the differentiated positions of cities within global fields of power (Jonas et al., 2018; Raco, 2018).

6.3.1 *What's New? Historicizing Local Migration Policies*

Local migration policymaking is often presented in the literature as a relatively recent phenomenon. Until about two decades ago, issues such as 'integration' were primarily framed and managed at the national level. This argument has been reinforced by frequent references to a 'local turn' in Anglophone migration studies (Scholten & Penninx, 2016; Zapata-Barrero et al., 2017). The idea of a 'local turn' not only captures researchers' growing dissatisfaction with nation-state centric analyses of migration policy, but is also used to suggest a shift in policymaking itself. This latter assertion, however, risks presenting an inaccurate view of the past.

In actual fact, the urban arena has long played a role in influencing and testing national legislation. For example, housing discrimination in major British cities in the 1960s, which was widely publicized in John Rex and Robert Moore's 1967 study of the Sparkbrook area of Birmingham, was instrumental in shaping the 1968 Race Relations Act, which extended anti-discrimination legislation to housing, employment and public services (Moore, 2011). Cities, however, were not simply locked into national legal frameworks. Throughout the post-war period, municipal governments in northern and western European cities responded to the arrival and settlement of foreign-born migrants in different ways. In an important comparative study, Hans

Mahnig (2004) examines the evolution of migration policymaking in Paris, West Berlin and Zurich between the 1950s and 1990s. Throughout this period, the local authorities in Paris prioritized improving the substandard living conditions of migrant workers, even though key decisions regarding urban renewal and social housing in the French capital were largely in the hands of the central state. In Berlin, the city government's initial plans to restrict the settlement of Turkish guest workers and encourage ethnic German immigration were abandoned in the 1980s in favour of a more realistic 'living together' approach, which included the appointment of Germany's first local 'Commissioner for Foreigners' responsible for coordinating activities across city departments. Finally, in Zurich, proposals for limited enfranchisement of the city's southern European guest workers were repeatedly rejected in local referendums, although the city government did manage to adopt some inclusive measures, particularly in education, from the 1980s onwards. For all their differences, migration in the three cities only became politically salient when it was perceived as a public order problem, although by the 1980s city authorities had begun to recognize migrants as permanent members of the local population.

A crucial factor overlooked in Mahnig's study was the role of local governments in administering social protection programmes at a time when welfare states were massively expanding across much of Europe. The fact, for instance, that municipalities in Scandinavian countries had a relatively high degree of autonomy in managing welfare services meant that they played a more active role in shaping local pathways of migrant incorporation than was the case elsewhere. Historian Heidi Van Jønsson (2013) compares the approaches in Copenhagen and the neighbouring new town of Ishøj in the 1970s. While the Danish capital followed the national trend of including new arrivals into existing welfare arrangements, Ishøj chose to formulate its own migrant policy, which included the introduction of counselling services and language classes, as well as a controversial 10 per cent quota for foreigners in social housing (a policy copied from the Swedish city of Malmö). Moreover, it was precisely the lack of national guidelines during this period that 'left room for other political actors – at the municipal level – to set the agenda, propose solutions and promote visions for early Danish integration policy' (Jønsson, 2013: 594).

Rather than confirming a 'local turn', the history of post-war municipal policymaking (see also Hackett, 2015; Schrover, 2021) encourages us to think more carefully about the broader societal changes that have underpinned the transformation of local migration policies in recent decades. First, the intensification of neoliberal globalization, the reconfiguration of state power, and the proliferation of political actors have combined to increase the importance of the city as a scale of governance and have changed what can be achieved locally. For example, the dismantling of post-war welfare systems and the marketization of public services have reduced the redistributive functions of the local state but increased its role in regulating the provision of services by third parties (Blanco et al., 2014). Second, the increase and diversification of international migration across Europe since the 1990s has reinforced the centrality of cities in migratory processes. Contemporaneously, the consolidation of communities with migration backgrounds, the gradual acquisition of citizenship and the entry of members of minority groups into local politics have raised the pressure on local governments to take action on migration-related issues. Third, the changing demographics of cities (due not only to migration but also to other processes such as suburbanization and gentrification), together with generational shifts in socio-cultural attitudes and the restructuring of urban economies, have meant that 'diversity' has increasingly been identified, if not celebrated, as a defining characteristic of metropolitan life, and has subsequently moved to the centre of policy attention. Finally, in light of the above changes, there has been a growing interest on the part of supranational institutions in the local dimension of migration policymaking. For example, the Eurocities network began to champion the role of cities in migration policies in the 1980s (Borja, 1992) and later organized local policy cooperation and exchange across Europe (Caponio & Borkert, 2010). Such initiatives have further put cities in the spotlight and widened the potential scope for action.

6.3.2 Local Migration Policies Between 'Integration' and 'Diversity'

What does local migration policy mean in practice today? The answer to this question is not straightforward. There is a clear gap between policy discourses and proposals on the one hand and their implementation on the other (Czaika & de Haas, 2013; Schiller, 2017). As a starting point, local migration policies can be divided into two general areas: '*integration*' policies, which aim to facilitate the social, political and economic incorporation of migrants in cities, and '*diversity*' policies, which seek to manage and mobilize the multidimensional differences associated with migration. While there is inevitable overlap between these two domains, maintaining this distinction is helpful in thinking about the different objectives of migration governance at the local level.

The term 'integration' is highly contested in the field of migration studies. Among other things, it is seen to naturalize migrants as problematic others, to downplay existing inequalities within the host society and to be structured around normative end goals (Favell, 2022; Schinkel, 2018). Rather than treating integration as a theoretical concept, it can be more productively used to describe a set of governmental techniques and discourses (Hadj Abdou, 2019). Integration thus encompasses the conditions and modalities through which migrants access local services upon arrival in a locality. This may consist of pathways to existing arrangements (for example, through the provision of temporary documentation to access health care), targeted long-term support such as language lessons, classroom assistance and vocational training programmes, or exceptional measures such as the provision of accommodation and food for asylum seekers dispersed by central government. In political terms, it could mean facilitating new voter registration, setting up consultative forums or simplifying naturalization criteria. To increase the feasibility of integration policies, councils often employ cultural mediators, usually migrants themselves, to ensure communication between public institutions and people with limited knowledge of local languages.

The goal of integration policies is never simply to improve the lives of migrants. Local governments are driven as much by pragmatic and self-interested concerns as by egalitarian principles: preparing newcomers to fill shortages in low-skilled sectors of the labour market, containing the potential spread of infectious diseases, mitigating social conflict, building alliances with minority communities, and so on. Moreover, integration is never framed solely in terms of people's needs but also according to less measurable criteria such as deservingness and personal responsibility. These are expressed in terms of newcomers' demonstrated willingness to acquire language proficiency, to interact with 'natives', to respect local values and norms, and to achieve social and economic autonomy. Such goals reflect the growing influence of state-led 'civic integration' and workfare programmes, particularly in northern European countries, where socio-economic integration is increasingly intertwined with the inculcation of liberal democratic values and neoliberal entrepreneurship of the self (Suvarierol, 2015). This has led scholars to question the scope for local policymakers to formulate alternative approaches to national integration strategies (Gebhardt, 2016).

'Diversity', like 'integration', does not have a fixed meaning in relation to migration. In some local policy arenas, diversity functions as a euphemism for the various challenges posed by migration, in others it replaces the contested paradigm of multiculturalism by encompassing a wider range of differences such as physical ability and gender (Schiller, 2017), while in some linguistic/national contexts it is rarely used at all (Escafré-Dublet & Lelévrier, 2019). Here, I use the term to refer to the sum of policy instruments used to govern migration-related differences, regardless of how these differences are locally recognized and classified.[1]

If integration policies primarily respond to the needs of newcomers, diversity policies are more ambivalent in that they tend to target both newcomers and established residents with a migration background. They may include measures to accommodate the cultural and religious practices of minorities, such as recognising and supporting associations, places of worship and educational establishments. Policies may extend to the setting of diversity quotas in the municipal workforce, for example through the employment of cultural mediators as mentioned above. In addition, diversity concerns spill over into other policy domains such as urban planning, where tasks can range from incorporating different cultural elements into the design of public spaces to developing neighbourhood facilities aimed at strengthening intergroup relations (Pemberton, 2017).

The accommodation of cultural diversity in cities is not simply a political or practical concern, but has often been mobilized for economic ends. The promotion of particular constellations of diversity, alongside localist claims to tolerance and openness, are increasingly seen as essential ingredients for global economic competitiveness and urban growth (Raco, 2018). Such a trajectory resonates, for example, with Richard Florida's (2004) thesis of the rise of a creative class based on the intra-urban mobility of cultural talent that captivated policymakers in European cities in

---

[1] For a comparative study of 'intercultural' policy programmes in European cities, see Hadj Abdou and Geddes (2017).

the early 2000s (Peck, 2012). The instrumental and selective use of migration-driven diversity in urban policy has been roundly criticized for failing to address cultural power relations in cities and for depoliticizing, if not exacerbating, existing social and economic inequalities (Hadj Abdou, 2019). Nonetheless, diversity has persisted as an ideological cornerstone of urban policy in many large European cities, shaping and legitimizing the ways in which migration is subsequently governed at the city scale (Raco & Kesten, 2018). This is not to deny the possibility of alternative policy approaches, but rather to point out that the valorization of 'diversity' in the context of neoliberal urban capitalism is never a neutral process.

6.3.3 Recent Research on Local Migration Governance

The last two decades have seen a veritable explosion of interest in the local dimension of migration policymaking. In particular, political scientists and policy analysts have produced an extensive body of empirical and theoretical work analyzing, among other things, the convergences and divergences between national and local policies, the different administrative capacities of local states, and the interactions and tensions between governmental and non-governmental actors (Penninx et al., 2004; Alexander, 2007; Caponio & Borkert, 2010; Jørgensen, 2012; Caponio et al., 2019). Some scholars have observed a growing trend towards the 'mainstreaming' of migration integration governance in a number of cities across Europe, where measures targeting specific groups have been abandoned in favour of area-based approaches that cut across various policy sectors (Scholten & van Breugel, 2018), while others have shown how local multiculturalist approaches have been adapted in the wake of the 'multiculturalism backlash' at the national level (Ambrosini, 2017). Despite the general recognition of the multilevel nature of contemporary migration governance, some researchers have emphasized the continuing decentralization of policy development (Caponio & Borkert, 2010), while others have pointed to examples of recentralization, with national governments 'returning' to steer policy agendas at the local level (Emilsson, 2015). There has also been growing interest in various subtopics that are overlooked or only cursorily addressed in general analyses, such as intercity policy networks and their impact on multilevel governance processes (Caponio, 2019; Flamant et al., 2022; Lacroix & Spencer, 2022), and the strategic opportunities and limits of key political actors such as mayors to lead on migration issues (Bazurli et al., 2022).

In order to make sense of the sheer diversity of local migration policy environments, many studies have drawn attention to the interplay of a wide range of contextual factors. Relevant factors include the economic power and size of cities (Çağlar & Glick Schiller, 2018; van Breugel, 2020), the type of institutional setting, such as the existence of directly elected mayoral systems (Bazurli et al., 2022), the nature of city-state relations and the degree of decentralization of decision-making (Alexander, 2007), the local political milieu and voting behaviour (de Graauw & Vermeulen, 2016), and the particular history and forms of migration in a given locality

(Dekker et al., 2015). Many studies are based on international comparative research, but some scholars have sought to unpack the incongruities between cities in a single country. For example, in her analysis of three major French cities (Nantes, Lyon and Strasbourg), Anouk Flamant (2020) identifies four factors to explain their different preferred approaches: the relationship with national authorities, the mobilization of European Union resources, the capacities of civil society and the career paths of policy officers.

This shared emphasis on contextual diversity has inadvertently highlighted the uneven production of knowledge within local policy research. Despite a steady increase in EU-funded pan-European research projects since the 2010s, the literature on local migration policy continues to be dominated by studies of cities in 'old' destination countries in northern and western Europe (Schmiz et al., 2022). This has set the frames of comparison and sometimes led to assumptions about the wider relevance of contingent processes. Research on 'new' urban destinations in the Mediterranean region has underscored crucial dimensions that are not always fully acknowledged in policy debates across Europe, such as the different historical and socio-economic conditions in which international migration emerged from the 1980s onwards (Dines, 2022), and the distinct welfare and housing regimes and trajectories of post-war urban development (Arbaci, 2019). The conspicuous lack of detailed research on urban governance in Central and Eastern Europe (CEE) has been attributed to the highly centralized nature of migration policy and the lower levels of international migration in the region's cities (Duszczyk et al., 2019). Since the late 2010s, however, the situation has started to change, with a move to include CEE cases in comparative projects and to explore how policy approaches circulate and adapt in new contexts (Scholten & van Breugel, 2018). Greater sensitivity to different urban contexts across Europe opens up the possibility of considering Mediterranean and CEE cities not in a perpetual game of catch-up with their northern and western counterparts, but as starting points for comparison. A case in point is the impact of the Russian invasion of Ukraine, which saw Warsaw receive over 500,000 arrivals in early 2022. This forced the city government to abandon its previous laissez-faire approach to Ukrainian circular migration and to implement an emergency integration policy to coordinate the large number of volunteers and provide support services (Wach & Pachocka, 2022).

Over the past decade, most political science research on local migration governance has focused on contextualizing, categorizing and evaluating policy discourses and practices, often within the confines of institutional settings, and has generally been less interested in bringing this analysis into conversation with broader discussions of contemporary urban society or migration. By contrast, geographers have tended to pay more attention to the connections between migration policies and urban policy agendas. For example, Mike Raco and Tuna Tasan-Kok have explored the different framings of diversity in a number of European cities and how these are incorporated into local development strategies (Raco, 2018; Raco & Tasan-Kok, 2019), while Jonathan Darling has examined the urban configuration of national asylum policy in the UK and how cities are implicated in the politicization of forced migration (Darling, 2017, 2021). Meanwhile, a separate and smaller body of ethnographic work by anthropologists and sociologists has investigated how local urban policies affect the livelihoods of migrants in cities and how migrants themselves are entangled in processes of urban regeneration (Çağlar & Glick Schiller, 2018; Dines, 2012). Although there is limited dialogue between different disciplinary approaches, this extensive and expanding body of literature provides important insights into the complexities of migration governance at the urban scale. That said, much of the empirical evidence in the policy literature is based on document analysis and interviews with policymakers. A lack of in-depth, embedded research means that we still know relatively little about the messy day-to-day processes of policymaking and the (unintended) consequences of policy actions.

6.4 Cities of Refuge, Sanctuary and Solidarity: Local Policy Actors Take a Stand

Much of local migration policymaking is routine and unremarkable. Sometimes it can become more visible, for example when local authorities decide to promote migration-driven diversity or, conversely, when they design policies that deliberately work against migrant incorporation. On occasions, however, visibility is actively sought. This has been particularly the case over the last decade, as an increasing number of mayors and local councils have publicly committed themselves to supporting undocumented migrants and asylum seekers and have devised means of countering hostile and/or inadequate national policies.

The nomenclature and content of such initiatives vary across national and local contexts and include a range of actions, from concrete measures to symbolic policies and advocacy for changes in government legislation. In Anglophone countries, the term 'sanctuary city' is commonly used by political actors and academics alike. In the US, where the term first emerged in San Francisco in the 1980s, the idea of urban sanctuary is mainly directed at limiting local cooperation with federal immigration policy such as waiving checks on a person's legal status, while in the UK, where a 'City of Sanctuary' movement was first endorsed by Sheffield City Council in 2007, the focus is instead on promoting values of hospitality and changing public attitudes towards refugees (Bauder, 2017). In southern European cities, other terminologies such as 'refuge' and 'solidarity' have been used. Following the increase in Mediterranean crossings in the 2010s, mayors and other municipal officials in frontline cities such as Palermo and Barcelona have challenged the rise of anti-immigrant populism and ineffective reception and relocation mechanisms with ad hoc policies such as registering new arrivals, enabling access to municipal services and establishing reception facilities that operate independently of the central state (Bazurli, 2019). The same actors have also actively engaged with like-minded municipalities in translocal networks to advocate for alternative approaches to migration management. This includes the Palermo Charter process, which has sought to create a series of 'open harbours' across Europe, inspired by the eponymous manifesto signed by the mayor of Palermo and others in 2015, which declared international mobility an inalienable human right (Ataç et al., 2021).

Sanctuary policies do not simply reflect the ideological positions of their proponents. Rather, they are often a response to practical concerns, such as meeting the urgent needs of undocumented residents or exerting pressure on national governments, and can also be a means of enhancing a city's reputation as 'tolerant' and 'cosmopolitan' or boosting the political capital of local leaders in the face of unpopular national rhetoric. Moreover, the development of a particular policy position is never the work of an enlightened administration alone, but is influenced by pressure and cooperation from non-state actors in social movements and civil society organizations. Strategic alliances, often the result of complex negotiations and compromises that reflect as much the steering capacity of migrant rights groups as the activism of municipalities, are crucial in shaping and securing political gains and public support. Similarly, the diplomatic channels linking sanctuary cities tend to differ markedly from mainstream policy networks in that they involve a greater mix of institutional and non-institutional participants.

Sanctuary cities have rapidly become the focus of a burgeoning interdisciplinary literature, ranging from legal and philosophical debates to analyses of the contingent processes that influence sanctuary policies and their outcomes in different cities, and which, in contrast to a lot of the research on local policymaking, pays greater attention to the urban contexts of governance and migration (for reviews, see Darling & Bauder, 2019; Paquet, 2023). While much of this research is sympathetic to the aims of sanctuary cities, it is by no means uncritical. One particularly powerful critique, with specific reference to UK cities, is that sanctuary policies operate to normalize the precarious situations of asylum seekers and to reinforce a state of limbo by prolonging their waiting time (Bagelman, 2016). This directly points to the limitations of local policies, such as the inescapable fact that city governments cannot grant asylum or prevent the state from detaining and removing people without status from their jurisdiction. Despite Barcelona's ambitious Refugee City programme launched by Mayor Ada Colau in 2015, which included an independent reception system and the promotion of civil society participation in policymaking, the city struggled to cope with increasing numbers of asylum seekers, while a preliminary agreement with the mayor of Athens to resettle 100 migrants in Barcelona was blocked by the Spanish government (Bazurli, 2019). It is also important to bear in mind that sanctuary projects can end abruptly following a change in leadership or an electoral defeat, as occurred in Palermo in 2022 and Barcelona in 2023. Nevertheless, sitting at the intersection of local policy and urban social movements, sanctuary cities have animated alternative courses of municipal action on migration issues and created political repertoires that can be developed, adapted or contested in other local or transnational contexts.

6.5 Migrant Politics in Cities

Migrants are and have always been political subjects, regardless of their legal status, length of residence or possession of the right to vote, even if these variables play an important role in determining the conditions, forms and goals of their political action. The spectrum of migrants' political action is vast. To take just the issues discussed above, it might include asylum seekers' struggles against long waiting times or their participation in sanctuary city movements; resident migrants' demands for cultural recognition or their contributions to policymaking as elected officials. At the same time, it should not be forgotten that people categorized as 'migrants' may also mobilize on the basis of other identities and affinities, and around issues that have nothing to do with migration.

The city represents a key context for migrant politics. This is not surprising given the ever-increasing concentration of migrants in urban areas and the growing role of local government in migration policy. In order to unpack the significance of the urban scale, however, we need to move beyond the assumption that the city is merely a setting for events and begin to see it as a particular *generative* space. Starting from this premise, geographers Walter Nicholls and Justus Uitermark attempt to theorize the relationship between migrant politics and cities. They highlight three urban processes that influence the formation of what they call 'immigrant counterpublics' (Nicholls & Uitermark, 2016: 880–885).[2] First, the large and heterogeneous populations of cities enable the development of economies of scale that sustain group-specific institutions such as social venues, civic associations and religious organizations, thereby facilitating the construction of parallel social, cultural and political worlds. Second, shared everyday experiences of hostility from dominant groups and discrimination in labour and housing markets can foster solidarity, strengthen collective identities and sharpen the oppositional edge of migrant groups. Finally, cities offer a wide range of activist networks, often with organizational skills and greater access to resources, that can guide political strategy and flank migrant mobilizations.

Nicholls and Uitermark's conceptualization of 'immigrant counterpublics' develops from their research on migrant rights activism in Amsterdam, Paris and Los Angeles (2017), and so may not be entirely relevant to smaller urban localities or specific groups of migrants, such as dispersed asylum seekers. Nevertheless, the three processes point to some essential political building blocks: the co-presence of difference, parallel institutions and coalition possibilities. There are other generative forces to consider. First, the growing interconnectedness of cities, reinforced by migrants' own transnational ties, shapes political practices and networks and amplifies their resonance across localities (Østergaard-Nielsen, 2003). Second, urban public spaces play a crucial role—alongside and often in conjunction with digital media—as material arenas for mediating political encounters, organizing actions,

[2] The idea of 'immigrant counterpublic' develops Nancy Fraser's concept of 'counterpublics', which Fraser defines as 'parallel discursive arenas where members of subordinated social groups invent and circulate counter discourses [excluded from the dominant public sphere] to formulate oppositional interpretations of their identities, interests and needs' (Nicholls & Uitermark, 2016: 879).

boosting visibility and creating durable infrastructures. For example, the 'Lampedusa in Hamburg' movement, consisting of over 300 non-status refugees who had arrived from Libya via Italy, set up a tent at a busy intersection in central Hamburg from 2013 until it was removed during the Covid-19 pandemic in 2020 to draw attention to their struggle for permanent residency and the right to work, and to provide an information exchange point for activists.

Migrant mobilizations in urban contexts vary over time and therefore need to be situated historically. To this end, Davide Però and John Solomos have identified three main moments in post-war migrant politics in Western Europe that can usefully be extended to think about cities (Però & Solomos, 2010: 2–7). They sketch an initial 'era of collective action' during the 1960s and 1970s in which international (and internal) labour migrants, when conditions allowed, joined local workers' organizations in struggles for material justice, but also began to self-organize around other issues such as racism and cultural recognition. From the late 1970s, against the rise of service economies and a retreat from class-based redistributive politics, ethnicity and cultural rights became increasingly prominent instruments for promoting migrant political participation inside and outside institutional settings. Since the 1990s, with the diversification and globalization of migration flows, coupled with increasingly restrictive immigration policies and economic deregulation, there has been an upsurge in grassroots political action around new issues such as regularization, precarious work and detention. Clearly, these three 'stages' cannot fully account for the range of experiences in any one city. Rather, they provide a basic historical framework that encourages us to remain alert to the changing conditions and articulations of politics.

Today, migrant politics in European cities are highly diverse. They address a panoply of migration-related issues—citizenship, racism, labour, cultural rights, etc.—but can also encompass broader issues such as the quality and distribution of public services. They take place within, outside and against local institutions and target different levels of government (Mamadouh, 2018: 469–472). The salience of particular demands, practices and alliances is influenced by, among other things, the socio-economic circumstances, migration histories and political opportunities in different cities. For instance, the migrant squatting movements unique to southern European cities such as Rome and Athens reflect the acute housing crises, migrant incorporation trajectories and social movement traditions in these localities (Montagna & Grazioli, 2019). Similarly, the composition and direction of political action may vary according to migrants' backgrounds, identities and legal status, their ideological values and particular grievances, their levels of attachment and engagement with different localities, and their relationships with other migrant groups, political actors and institutions.

Scholarship on migrant politics in urban contexts has long been dominated by an interest in migrants' political participation in local institutions. Early studies drew on political opportunity structure theory from social movement studies (Ireland, 1994) and adopted historical institutional approaches to analyze variation across cities (Garbaye, 2006), while later research, often based on survey data, emphasized the importance of social capital and political trust generated, for example, through migrant associations (Jacobs & Tillie, 2004; Morales & Giugni, 2011). From the

2000s onwards, a body of work by anthropologists and sociologists on migrants' rights and labour politics began to challenge the previous emphasis on ethnicity to explain local political participation, highlighting instead more complex constellations of factors such as people's migratory projects, their transnational networks and the conditions they encountered in destination cities (Però, 2008). Over the past decade, and in contrast to scholarship on local policymaking, there has been a notable increase in fine-grained ethnographic research exploring lived experiences of different forms of mobilization. For example, Kari Anne Drangsland (2020) examines the conflicted responses of members of Lampedusa in Hamburg to the state government's offer of conditional legalization, which fell short of collective demands. More recently, activist scholars explicitly aligned with migrants' struggles have advanced thinking about migrant political subjectivities in relation to the urban arena by opening up conversations around ideas of resistance, reproductive politics and racialization, although much of this work focuses on specific types of migrants, particularly those arriving via irregular routes (Kapsali, 2020; Santamarina, 2024).

Migrant politics—whether migrant-led, alongside other political actors, or through local institutions—redefine the political landscapes of cities. Struggles over the distribution of rights and resources, for example, raise broader questions about the organization of cities themselves. In the process, migrants may upscale their demands to higher levels of government, as in the case of the Syrian hunger strikers in Athens, and at the same time rescale the politics of migration, exemplified by Lampedusa in Hamburg, where the external borders of the European Union are brought into the heart of a northern German city.

6.6 The Urban Scale of Anti-migration Politics

Cities have historically been central to the production of anti-immigration sentiments and politics. A case in point is Enoch Powell's 1968 'Rivers of Blood' speech, in which the English Conservative politician railed against non-white immigration and anti-discrimination legislation. Widely regarded as the most notorious speech in post-war British political history, Powell's words continue to haunt the country's conversation about migration and racism. What is often forgotten, however, is that Powell's invective was constructed around what he believed was happening in the West Midlands city of Wolverhampton. It was in Wolverhampton that 'respectable street[s]' inhabited by 'ordinary, decent, sensible people' were being 'taken over' by 'Negroes' and turned into 'place[s] of noise and confusion,' and it was here that local Sikh bus drivers had decided to 'agitate and campaign against their fellow citizens' for the right to wear turbans at work (Powell, 1968). Powell contrasts the chaotic present and a potentially ominous future with a vision of a socially harmonious and racially homogeneous recent past, a bygone era that omits any mention of Wolverhampton's long history of labour migration and industrial conflict (Hirsch, 2018). By speculating on the conditions in one particular city, Powell raises the stakes to call for a nationwide halt to Commonwealth immigration.

Powell articulated a series of racialized discourses about migrants in cities—nostalgia for a pristine urban world, the impossibility of coexistence, the erosion of local ways of life, the native community as 'a persecuted minority' (Powell, 1968)—that have since become familiar fodder for critics of immigration. In the decades that followed, such tropes were recoded in different contexts and amplified by a range of political actors, from neo-fascist parties such as the National Front in the UK, which violently targeted migrant and ethnic minority populations in cities such as Wolverhampton in the 1970s (Husbands, 1983), to the middle-class neighbourhood committees of Italian cities in the 1990s, which organized night patrols to deter public gatherings and unruly behaviour by local migrant residents (Dines, 2012). When not blatantly racist or scapegoating migrants for structural problems like unemployment (as in the former case), actors tended to frame their nativist claims in universalizing terms: theirs was a campaign for the common good in defence of public order and urban decorum. At the same time, these various manifestations of organized anti-immigrant sentiment need to be understood in the context of shifting local and national politics, socio-economic conditions and migration dynamics. For example, the National Front's rapid rise to become Britain's fourth-largest party in mid-1970s occurred against a backdrop of austerity politics, the normalization of racist tropes in mainstream media and politics, and public concerns about South Asian immigration to Britain fuelled by Enoch Powell's speech. The party's equally swift downfall during the latter half of the decade was precipitated by direct opposition from a mass anti-fascist movement and by the loss of its main mobilizing points to the new Conservative national government elected in 1979, which promised to take a tougher stance on immigration (Copsey, 2008; Townsend, 2017). The anti-migrant neighbourhood committees in Italian cities emerged during the 1990s in response to the settlement of international migrants in cheap, non-gentrified parts of city centres, at a time when political and media debates about immigration were closely linked with wider discourses about 'urban insecurity' (Foot, 2001). The visibility and influence of these committees waned considerably after two major immigration laws in 1998 and 2001 responded to their demands for institutional action, and following a shift in attention of the national media's coverage of immigration away from the urban sphere and towards external borders such as Lampedusa (Dines et al., 2016).

Anti-immigrant attitudes have also percolated into local policymaking. Until the 1980s, most city governments in western Europe, regardless of ideological persuasion, addressed immigration only when it became an unavoidable public order issue, and were more likely to promote exclusionary measures such as settlement bans than to argue for pragmatism or greater openness (Mahnig, 2004). Despite the spread of more accommodating approaches at the urban scale, local policies can still work against the interests of migrants. Over the last two decades, Italian municipalities have introduced a raft of exclusionary measures such as anti-begging decrees and bans on kebab shops and mosques (Ambrosini, 2013) and have refused to cooperate with national refugee distribution programmes (Marchetti, 2020). Such policies are most common in small and medium-sized cities and among right-wing administrations seeking to build political consensus around 'locals first' platforms. Although some proposals are deliberately designed to provoke a public response, some measures may be subtle and insidious, such as the introduction of stricter eligibility criteria for benefits, while others may be embedded in nominally inclusive agendas such as regeneration schemes in migrant neighbourhoods that discourage the arrival of low-income newcomers and threaten to remove key social infrastructures (Dines, 2009).

Local anti-immigration policies do not always achieve their desired goals: some measures simply prove unworkable and many face public opposition in the form of organized protests, negative media and legal challenges, and can be forcibly removed by the state if declared unlawful (Ambrosini, 2013). Moreover, aspects over which municipalities have little control—support networks and labour and housing markets—tend to be more influential in people's decisions to move to a particular locality. Nevertheless, regardless of their operational effectiveness, exclusionary policies can have serious negative consequences insofar as they exacerbate the social and symbolic marginalization of migrants from local life. They can also reinforce discriminatory attitudes among local residents and lay the groundwork for approaches adopted elsewhere, including at higher levels of government, as in the case of the Danish state's 'ghetto laws' (Sect. 4.3.4), which drew inspiration from the aforementioned local quota experiments of the 1970s and 1980s (Jønsson, 2013).

Despite its longstanding entanglement with the urban scale, anti-immigration politics continues to be largely associated with the nation-state. This has been compounded by the rise of far-right populist movements across Europe and beyond, which have gained a foothold in mainstream politics since the 2008 economic crisis and have contributed to the intense politicization of migration. According to liberal commentators, these pre-eminently national(ist) phenomena are the antithesis of the inclusive democratic processes augured by cities (Muggah & Glenny, 2017). However, these forces have much more to do with the local level than is often acknowledged. Some national organizations, such as the Lega in Italy, have their origins in local and regional politics, while others were founded specifically as urban movements. PEGIDA (Patriotic Europeans against the Islamization of the West), for example, was founded in Dresden and organized weekly demonstrations against Islam and immigration that were replicated in cities and towns across Germany and abroad. The fact that PEGIDA attracted the largest crowds and persisted the longest in Dresden has been attributed to the movement's ability to tap into the hegemonic reactionary political culture of the city and surrounding region and to mimic local urban protest traditions (Virchow, 2016). These nativist exploits cannot be explained away as the exceptional experiences of second-tier post-socialist cities. In fact, Dresden also attracted some of the largest counter-demonstrations in Germany. On the contrary, the organizational nerve centres of xenophobic populism are very often located in major cities; cities that are routinely conceived as global and diverse, and therefore presumably immune to such politics. In her analysis of the spatial politics of VOX in a Madrid suburb, Ana Santamarina considers how the Spanish far-right party stoked tensions around a reception centre for foreign minors in an attempt to transform its national anti-immigration platform into a set of concerns appealing to 'ordinary people' in everyday urban settings (Santamarina, 2021: 897). Stressing the need to move beyond nation-centred and institutional accounts of parties like VOX,

Santamarina calls for a greater focus on the situated inequalities and grievances at the neighbourhood level that enable the social and political reproduction of far-right populism, but also the articulation of forms of resistance. Ultimately, this means bringing the methodological nationalism critique to bear *also* on our understandings and imaginaries of anti-immigration politics.

6.7 Conclusion

This chapter has considered the policies and politics that emerge out of the relationship between migration and cities. Drawing on literature from a range of disciplines, it has examined the ways in which pro- and anti-migration policies and politics change over time and across different localities, and how these changes relate to various sources of power and influence operating at different scales, including the nation-state. The city has been understood both as a political space that generates opportunities and possibilities and as a focus for intervention and potential transformation. The seemingly distant worlds of city enthusiast Khalid Koser and the Syrian hunger strikers yearning to leave Athens, provocatively juxtaposed at the beginning of the chapter, ultimately encourage us to reflect on the possible intersections between policy discourse on the one hand and migrant agency and collective action on the other. In doing so, we are better able to comprehend the complexities of how urban government is both enacted and contested and to remain alert to the unpredictable effects and politics that are produced (Darling, 2021). Bringing migration policy and migrant politics into the same discussion can help to avert analytical blind spots, such as overplaying the propensity of urbanites to embrace diversity or conflating the urban and the local, and to foreground broader issues such as the unequal distribution of resources.

References

Alexander, M. (2007). *Cities and labour immigration: Comparing policy responses in Amsterdam, Paris, Rome and Tel Aviv*. Ashgate.

Ambrosini, M. (2013). "We are against a multi-ethnic society": Politics of exclusion at the urban level in Italy. *Ethnic and Racial Studies, 36*(1), 136–155.

Ambrosini, M. (2017). Superdiversity, multiculturalism and local policies: A study on European cities. *Policy and Politics, 45*(4), 585–603.

Arbaci, S. (2019). *Paradoxes of segregation: Housing systems, welfare regimes and ethnic residential change in southern European Cities*. Wiley.

Ataç, I., Rygiel, K., & Stierl, M. (2021). Building transversal solidarities in European cities: Open harbours, safe communities home. *Critical Sociology, 47*(6), 923–939.

Bagelman, J. (2016). *Sanctuary city: A suspended state*. Palgrave Macmillan.

Bauböck, R. (2003). Reinventing urban citizenship. *Citizenship Studies, 7*(3), 139–160.

Bauböck, R., & Orgad, L. (Eds.). (2020). *Cities vs States: Should urban citizenship be emancipated from nationality?* European University Institute.

Bauder, H. (2016). Possibilities of urban belonging. *Antipode, 48*(2), 252–271.
Bauder, H. (2017). Sanctuary cities: Policies and practices in international perspective. *International Migration, 55*(2), 174–187.
Bazurli, R. (2019). Local governments and social movements in the "refugee crisis": Milan and Barcelona as "Cities of Welcome." *South European Society and Politics, 24*(3), 343–370.
Bazurli, R., Caponio, T., & de Graauw, E. (2022). Between a rock and hard place: Mayors, migration challenges and mutlilevel political dynamics. *Territory, Politics, Governance, 10*(3), 297–305.
Blanco, I., Griggs, S., & Sullivan, H. (2014). Situating the local in the neoliberalisation and transformation of urban governance. *Urban Studies, 51*(15), 3129–3146.
Borja, J. (1992). Eurocities: A system of major urban centers in Europe. *Ekistics, 59*(352/353), 21–27.
Çağlar, A., & Glick Schiller, N. (2018). *Migrants and city-making: Dispossession, displacement and urban regeneration.* Duke University Press.
Caponio, T. (2019). City networks and the multilevel governance of migration: Towards a research agenda. In T. Caponio, P. Scholten, & R. Zapata-Barrero (Eds.), *The Routledge handbook of the governance of migration and diversity in cities* (pp. 182–192). Routledge.
Caponio, T., & Borkert, M. (Eds.). (2010). *The local dimension of migration policymaking.* Amsterdam University Press.
Caponio, T., & Jones-Correa, M. (2018). Theorising migration policy in multilevel states: The multilevel governance perspective. *Journal of Ethnic and Migration Studies, 44*(12), 1995–2010.
Caponio, T., Scholten, P. & Zapata-Barrero, R. (Eds.). (2019). *The Routledge handbook of the governance of migration and diversity in cities.* Routledge.
Copsey, N. (2008). *Contemporary British fascism: The British National Party and the quest for legitimacy.* Palgrave Macmillan.
Czaika, M., & De Haas, H. (2013). The effectiveness of immigration policies. *Population and Development Review, 39*(3), 487–508.
Darling, J. (2017). Forced migration and the city: Irregularity, informality, and the politics of presence. *Progress in Human Geography, 41*(2), 178–198.
Darling, J. (2021). Refugee urbanism: Seeing asylum "like a city." *Urban Geography, 42*(7), 894–914.
Darling, J., & Bauder, H. (Eds.). (2019). *Sanctuary cities and urban struggles: Rescaling migration, citizenship, and rights.* Manchester University Press.
De Graauw, E., & Vermeulen, F. (2016). Cities and the politics of immigrant integration: A comparison of Berlin, Amsterdam, New York City, and San Francisco. *Journal of Ethnic and Migration Studies, 42*(6), 989–1012.
Dekker, R., Emilsson, H., Krieger, B., & Scholten, P. (2015). A local dimension of integration policies? A comparative study of Berlin, Malmö, and Rotterdam. *International Migration Review, 49*(3), 633–658.
De Shalit, A. (2018). *Cities and immigration: Political and moral dilemmas in the new era of migration.* Oxford University Press.
Dikeç, M., & Swyngedouw, E. (2017). Theorizing the politicizing city. *International Journal of Urban and Regional Research, 41*(1), 1–18.
Dimitriadis, I., Hajer, M. H., Fontanari, E., & Ambrosini, M. (2021). Local "battlegrounds." Relocating multi-level and multi-actor governance of immigration. *Revue Européenne des Migrations Internationales, 37*(1), 251–275.
Dines, N. (2009). The disputed place of ethnic diversity: An ethnography of the redevelopment of a street market in East London. In R. Imrie, L. Lees, & M. Raco (Eds.), *Regenerating London: Governance, sustainability and community in a global city* (pp. 254–272). Routledge.
Dines, N. (2012). *Tuff city: Urban change and contested space in central Naples.* Berghahn.
Dines, N. (2022). How far can we compare? Migration studies, comparative urbanism and the potential of a trans-Mediterranean perspective. *International Migration, 60*(1), 205–218.
Dines, N. (2023). After entry: Humanitarian exploitation and migrant labour in the fields of southern Italy. *Environment and Planning d: Society and Space, 41*(1), 74–91.

Dines, N., Montagna, N., & Ruggiero, V. (2016). Thinking Lampedusa: Border construction, the spectacle of bare life and the productivity of migrants. *Ethnic and Racial Studies, 38*(3), 430–445.

Drangsland, K. A. T. (2020). Waiting as a redemptive state: "The Lampedusa in Hamburg" and the offer from the Hamburg government. *Time & Society, 29*(2), 318–339.

Duszczyk, M., Pszczółkowska, D., & Wach, D. (2019). Warsaw: A new immigration city in search of its integration policy towards newcomers. In T. Caponio, P. Scholten, & R. Zapata-Barrero (Eds.), *The Routledge handbook of the governance of migration and diversity in cities* (pp. 130–142). Routledge.

Emilsson, H. (2015). A national turn of local integration policy: Multi-level governance dynamics in Denmark and Sweden. *Comparative Migration Studies, 3*(7). https://doi.org/10.1186/s40878-015-0008-5

Escafré-Dublet, A., & Lelévrier, C. (2019). Governing diversity without naming it: An analysis of neighbourhood policies in Paris. *European Urban and Regional Studies, 26*(3), 283–296.

Favell, A. (2022). *The integration nation: Immigration and colonial power in liberal democracies.* Polity Press.

Filomeno, F. A. (2017). *Theories of local immigration policy.* Palgrave Macmillan.

Flamant, A. (2020). The local turn in integration policies: Why French cities differ. *Ethnic and Racial Studies, 43*(11), 1981–2000.

Flamant, A., Fourot, A. C., & Healy, A. (2022). Special issue: City network activism and the governance of migration. *Local Government Studies, 48*(6), 1017–1026.

Florida, R. (2004). *Cities and the creative class.* Routledge.

Foot, J. (2001). San Salvario, Turin: The creation of a dangerous place 1990–99. In R. King (Ed.), *The Mediterranean passage: Migration and new cultural encounters in southern Europe* (pp. 206–230). Liverpool University Press.

Garbaye, R. (2006). *Getting into local power: The politics of ethnic minorities in British and French cities.* Blackwell.

Garbin, D., & Millington, G. (2018). "Central London under siege": Diaspora, "race" and the right to the (global) city. *The Sociological Review, 66*(1), 138–154.

Gargiulo, E. (2017). The limits of local citizenship: Administrative borders within the Italian municipalities. *Citizenship Studies, 21*(3), 327–343.

Gebhardt, D. (2016). When the state takes over: Civic integration programmes and the role of cities in immigrant integration. *Journal of Ethnic and Migration Studies, 42*(5), 742–758.

Hackett, S. (2015). The "local turn" in historical perspective: Two city case studies in Britain and Germany. *International Review of Administrative Sciences, 83*(2), 340–357.

Hadj Abdou, L. (2019). Immigrant integration: The governance of ethno-cultural differences. *Comparative Migration Studies, 7*(15). https://doi.org/10.1186/s40878-019-0124-8

Hadj Abdou, L., & Geddes, A. (2017). Managing superdiversity? Examining the intercultural policy turn in Europe. *Policy & Politics, 45*(4), 493–510.

Harvey, D. (1989). From managerialism to entrepreneurialism: The transformation in urban governance in late capitalism. *Geografiska Annaler: Serie b, Human Geography, 71*(1), 3–17.

Harvey, D. (2012). *Rebel cities: From the right to the city to the urban revolution.* Verso.

Hirsch, S. (2018). *In the shadow of Enoch Powell: Race, locality and resistance.* Manchester University Press.

Hoekstra, M. (2015). Diverse cities and good citizenship: How local governments in the Netherlands recast national integration discourse. *Ethnic and Racial Studies, 38*(10), 1798–1814.

Husbands, C. T. (1983). *Racial exclusionism and the city: The urban support of the National Front.* Allen and Unwin.

Ireland, P. (1994). *The policy challenge of ethnic diversity: Immigrant politics in France and Switzerland.* Harvard University Press.

Isin, E. F., & Nielsen, G. M. (2008). *Acts of citizenship.* Zed Books.

Jacobs, D., & Tillie, J. (2004). Introduction: Social capital and political integration of migrants. *Journal of Ethnic and Migration Studies, 30*(3), 419–427.

Jønsson, H. V. (2013). Immigrant policy developing in Copenhagen and Ishøj in the 1970s. *Scandinavian Journal of History, 38*(5), 590–611.

Jonas, A. E. G., Miller, B., Ward, K., & Wilson, D. (2018). Spaces of urban politics: An introduction. In K. Ward, A. E. G. Jonas, B. Miller, & D. Wilson (Eds.), *The Routledge handbook on spaces of urban politics* (pp. 1–10). Routledge.

Jørgensen, M. B. (2012). The diverging logics of integration policy making at national and city level. *International Migration Review, 46*(1), 244–278.

Kaika, M., & Karaliotas, L. (2017). Athen's Syntagma Square reloaded: From staging disagreement towards instituting democratic spaces. In J. Hou & S. Knierbein (Eds.), *City unsilenced: Urban resistance and public space in the age of shrinking democracy* (pp. 121–132). Routledge.

Kapsali, M. (2020). Political infrastructures of care: Collective home making in refugee solidarity solidarity squats. *Radical Housing Journal, 2*(2), 13–24.

Kofman, E. (2024). Gendered citizenship and the right to the city. In L. Peake, A. Datta, & G. Adeniyi-Ogunyankin (Eds.), *Handbook on gender and cities* (pp. 162–171). Elgar.

Kohn, M. (2020). The danger zone: Charter cities, citizenship, and social justice. In R. Bauböck & L. Orgad (Eds.), *Cities vs states: Should urban citizenship be emancipated from nationality?* (pp. 46–48). European University Institute.

Koser, K. (2014). *Cities and the case for migration.* https://www.youtube.com/watch?v=rD7xxFCyPtQ

Kotronaki, L. (2018). Outside the doors: Refugee accommodation squats and heterotopy politics. *The South Atlantic Quarterly, 117*(3), 914–924.

Lacroix, T., & Spencer, S. (2022). City networks and the multi-level governance of migration. *Global Networks, 22*(3), 349–362.

Magnusson, W. (2011). *Politics of urbanism: Seeing like a city.* Routledge.

Mahnig, H. (2004). The politics of minority-majority relations: How immigrant policies developed in Paris, Berlin and Zurich. In R. Penninx, K. Kraal, M. Martiniello, & S. Vertovec (Eds.), *Citizenship in European cities: Immigrants, local politics and integration policies* (pp. 17–37). Ashgate.

Mamadouh, V. (2018). A city of migrants: Migration and urban identity politics. In K. Ward, A. E. G. Jonas, B. Miller, & D. Wilson (Eds.), *The Routledge handbook on spaces of urban politics* (pp. 468–478). Routledge.

Marchetti, C. (2020). Cities of exclusion: Are local authorities refusing asylum seekers? In M. Ambrosini, M. Cinalli, & D. Jacobson (Eds.), *Migration, borders and citizenship: Between policy and public spheres* (pp. 237–263). Palgrave Macmillan.

Margaronis, M. (2014, December 5). The anarchist and the refugees: Two hunger strikes in Athens. *The Nation.* https://www.thenation.com/article/archive/anarchist-and-refugees-two-hunger-strikes-athens/

Montagna, N., & Grazioli, M. (2019). Urban common and freedom of movement: The housing struggles of recently arrived migrants in Rome. *Citizenship Studies, 23*(6), 577–592.

Moore, R. (2011). Forty-four years of debate: The impact of race, community and conflict. *Sociological Research Online, 16*(3), 1–8.

Morales, L., & Giugni, M. (2011). *Social capital, political participation and migration in Europe: Making multicultural democracy work.* Palgrave Macmillan.

Muggah, R., & Glenny, M. (2017). *Populism is poison. Plural cities are the antidote.* https://www.weforum.org/agenda/2017/01/populism-is-poison-plural-cities-are-the-antidote/

Nicholls, W. J., & Uitermark, J. (2016). Migrant cities: Place, power, and voice in the era of super diversity. *Journal of Ethnic and Migration Studies, 42*(6), 877–892.

Nicholls, W. J., & Uitermark, J. (2017). *Cities and social movements: Immigrant rights activism in the United States, France, and the Netherlands, 1970–2015.* Blackwell.

Østergaard-Nielsen, E. (2003). The politics of migrants' transnational political practices. *International Migration Review, 37*(3), 760–786.

Paquet, M. (2023). Sanctuary cities. In A. Triandafyllidou (Ed.), *Routledge handbook of immigration and refugee studies* (pp. 434–443). Routledge.

Peck, J. (2012). Recreative city: Amsterdam, vehicular ideas and the adaptive spaces of creativity policy. *International Journal of Urban and Regional Research, 36*(3), 462–485.
Pemberton, S. (2017). Urban planning and the challenge of super-diversity. *Policy and Politics, 45*(4), 623–641.
Penninx, R., Kraal, K., Matiniello, M., & Vertovec, S. (Eds.). (2004). *Citizenship in European cities: Immigrants, local politics and integration policies*. Ashgate.
Però, D. (2008). Political engagement of Latin American in the UK: Issues, strategies, and the public debate. *Focaal, 51*(1), 75–90.
Però, D., & Solomos, J. (2010). Introduction: Migrant politics and mobilization: Exclusion, engagements, incorporation. *Ethnic and Racial Studies, 33*(1), 1–18.
Powell, E. (1968). *Enoch Powell's "Rivers of blood" speech*. https://anth1001.files.wordpress.com/2014/04/enoch-powell_speech.pdf
Purcell, M. (2006). Urban democracy and the local trap. *Urban Studies, 43*(11), 1921–1941.
Raco, M. (2018). Critical urban cosmopolitanism and the governance of urban diversity in European cities. *European Urban and Regional Studies, 25*(1), 8–23.
Raco, M., & Kesten, J. (2018). The politicisation of diversity planning in a global city: Lessons from London. *Urban Studies, 55*(4), 891–916.
Raco, M., & Tasan-Kok, T. (2019). Governing urban diversity: Multi-scalar representations, local contexts, dissonant narratives. *European Urban and Regional Studies, 26*(3), 230–238.
Russell, B. (2019). Beyond the local trap: New municipalism and the rise of fearless cities. *Antipode, 51*(3), 989–1010.
Santamarina, A. (2021). The spatial politics of far-right populism: VOX, anti-fascism and neighbourhood solidarity. *Critical Sociology, 47*(6), 891–905.
Santamarina, A. (2024). Racial capitalism, political reproduction, and the commons: Insights from migrant solidarity politics in Glasgow. *Antipode, 56*(1), 229–248.
Schiller, M. (2017). The implementation trap: The local level and diversity policies. *International Review of Administrative Sciences, 83*(2), 267–282.
Schinkel, W. (2018). Against "immigrant integration": For an end to neocolonial knowledge production. *Comparative Migration Studies*, *6*(31), https://doi.org/10.1186/s40878-018-0095-1
Schmiz, A., Felgentreff, C., Franz, M., Paul, M., Pott, A., Räuchle, C., & Schrader, S. (2022). Cities and migration: Bibliometric evidence from a spatially biased field of knowledge production. *Geographical Review, 112*(2), 267–285.
Scholten, P., & Penninx, R. (2016). The multilevel governance of migration and integration. In B. Garcés-Mascareñas & R. Penninx (Eds.), *Integration processes and policies in Europe: Contexts, levels and actors* (pp. 91–108). Springer.
Scholten, P., & van Breugel, I. (2018). *Mainstreaming integration governance: New trends in migrant integration policies in Europe*. Palgrave Macmillan.
Schrover, M. (2021). Rats, rooms and riots: Usage of space by immigrants in the Dutch Town Utrecht 1945–1970. *Journal of Migration History, 7*, 244–271.
Suvarierol, S. (2015). Creating citizen-workers through civic integration. *Journal of Social Policy, 44*(4), 707–727.
Thomaz, A. D. (2022). Urban citizenship for all? Exploring the limits of an agenda in São Paulo's squats. *Migration Studies, 10*(2), 214–234.
Townsend, M. (2017, 13 August). How the battle of Lewisham helped to halt the rise of Britain's far right. *The Guardian*. https://www.theguardian.com/uk-news/2017/aug/13/battle-of-lewisham-national-front-1977-far-right-london-police
van Breugel, I. (2020). Towards a typology of local migration diversity policies. *Comparative Migration Studies, 8*(23). https://doi.org/10.1186/s40878-020-00179-0
Virchow, F. (2016). PEGIDA: Understanding the emergence and essence of nativist protest in Dresden. *Journal of Intercultural Studies, 37*(6), 541–555.
Wach, D., & Pachocka, M. (2022). Polish cities and their experience in integration activities—The case of Warsaw. *Studia Europejskie, 26*(2), 89–105.

Zapata-Barrero, R., Caponio, T., & Scholten, P. (2017). Theorizing the "local turn" in a multi-level governance framework of analysis: A case study in immigrant policies. *International Review of Administrative Sciences, 83*(2), 241–246.

Chapter 7
Conclusion

7.1 Critical European Perspectives on the Migration-City Nexus

Over the last three decades, research on the relationship between migration and cities has developed as a multidimensional and multidisciplinary field of inquiry. It has sometimes converged around theoretical positions and common observations, but has often moved along parallel tracks with little cross-dialogue. This book has attempted to provide a guide to navigating the diverse literature on migration and European cities, and has explored key issues and concepts that emerge at their intersection. The five main chapters examine this relationship from historical, economic, socio-spatial, cultural and political/policy perspectives. The chapters share a common emphasis on the underlying conditions and processes that shape migrant incorporation, the ways in which themes and terminologies shift across the literature, and the need to attend to contextual diversity, not to make the obvious point that cities are different, but to grapple with the situated nature of knowledge production.

The book brings into comparative discussion the extensive research on different cities in north-western Europe, the theoretical heartland of European migration and urban studies, and the contributions of scholars working in the southern and eastern peripheries of the continent (Maloutas, 2012; Dines, 2016; Ha & Picker, 2022). Research agendas on topics as diverse as labour migration, entrepreneurship, segregation, cultural heritage and local policy can only be advanced if the implications of diverse contexts are brought to the fore. Sonia Arbaci's (2019) seminal study *Paradoxes of segregation* offers an illustrative lesson. The metaphors and indices of segregation developed in the United States and north-western Europe may provide points of reference for initiating research in southern European cities, but they are not conducive to interpreting processes shaped by different housing and welfare systems and alternative migration histories that lead to unpredictable outcomes, such as vertical stratification in central parts of Naples and Athens (Maloutas &

N. Dines, *Migration and European Cities*, IMISCOE Research Series,
https://doi.org/10.1007/978-3-032-15379-1_7

Karadimitriou, 2022; Sect. 4.3.3). Similarly, generic typologies such as ‘port’, ‘post-industrial’ or ‘medium-sized’ cities are not, in and of themselves, particularly helpful in comprehending the multi-layered nature of local contexts or the positions of cities in relation to the wider dynamics of national and global power. Moreover, they do not allow us to entertain the possibility that, say, past factory work or a maritime vocation may have negligible bearing on a city’s relationship with migration. Consider, for instance, the premise that viewing Naples as a ‘port city’ will be relevant to understanding the particular imprint of migration on this city (Pisarevskaya & Scholten, 2025: 1–2). Foregrounding this feature might serve to capture some symbolic aspect of the city’s central Mediterranean location. The fact of the matter, however, is that there has been no correlation between the port of Naples and international migration in the local area over the last fifty years. This is not only because the vast majority of migrants have not arrived in Naples by sea, but also because the port itself has held a somewhat marginal position in the city’s recent economic and labour history.

I have suggested that nexus thinking is useful in orienting our attention towards the various connections and interdependencies between migration and cities. The ‘migration-city nexus’ is proposed as an operational idea for conducting research and engaging with different literatures. It is not intended as a shorthand for the sum of all the various dimensions of the relationship between migration and cities. Rather, it encourages us to think carefully about the consequences of using different theories and typologies, and alerts us to the limitations of collapsing the relationship into a single overarching theory or relying on proxy terms such as ‘diversity’ to do the work for us. The ‘migration-city nexus’ is about migration *and* cities, but it is also about migration *in* and *to* cities. It is about making sense of migrants’ contributions to urban restructuring and rescaling, but also their embodied experiences and sense-making in the city as a particular *place*. It means taking seriously the pioneering theoretical work of key scholars such as Nina Glick Schiller and Ayse Çağlar, but also critically acknowledging the wealth of earlier research stretching back to the post-war period and beyond, which the same two authors too readily dismiss as obsolete (Glick Schiller & Çağlar, 2009: 180–186).

In other words, the ‘migration-city nexus’ can be usefully understood as a mnemonic device. When analysing phenomena or reviewing literature, it prompts us to think about what kinds of connections are being made, what concepts and discourses are being produced, and what is being foregrounded or overlooked. It prevents us from taking the migration-city relationship for granted and invites us to reflect on how we conceive migration and the city as our objects of analysis (Anderson, 2019; Dahinden, 2016; Kaufmann, 2012).

7.2 Political Ruptures, Global Pandemic and War

As interest in the migration-city nexus continues to grow, and as scholars seek to hone their analytical tools to comprehend convergences and divergences across settings, research will inevitably be pressed to address how the relationship between migration and cities evolves in the face of the long-term changes and abrupt shocks that are reshaping migration flows and urban environments across Europe and the rest of the world. Discussions of 'majority-minority' and 'no-majority' cities, where the old white majority group has become a minority, and the implications these have for revising ideas about 'assimilation' and 'segregation,' attest to the continuing role of migration and post-migration processes in shaping the demographic make-up of Europe's cities (Crul, 2016; Harris, 2024). These developments are simultaneously entangled with the ongoing restructuring of urban capitalism driven by, inter alia, the rise of platform and gig economies, the digitalization of urban systems, the financialization of housing and the intensification of urban tourism, which together reinforce the role of the city as a site of value extraction (Aalbers, 2018; Moisio & Rossi, 2024; Srnicek, 2017).

Migration research inevitably shifts its interests in response to the twists and turns of social change. However, as Loïc Wacquant reminds us, 'the pace of research is not that of media commentary or public action, and [...] the task of social science is not to surf the wave of current events, but to bring to light the durable mechanisms that produce them' (2007: 282). Moreover, drawing attention to the growing number of declared 'crises' around migration and cities in recent years—from the surge in trans-Mediterranean crossings since 2015 to the devastating impact of financialization on people's access to decent housing in major cities—also means unpacking the ways in which 'crisis' itself operates as a powerful narrative device that frames knowledge about social phenomena (Dines et al., 2018; Roitman, 2014). The Birmingham School classic *Policing the crisis* (Hall et al., 1978) masterfully shows us that the connections between cities, racism, immigration and 'crisis talk' are by no means new.

Nonetheless, a number of recent events have opened up interesting new avenues for research and have challenged some of the common assumptions about the links between migration and cities. The global rise of right-wing nationalist populism and the accompanying backlash against the mainstreaming of diversity politics and economic globalization more generally (Walter, 2021), has shattered the fiction that contemporary cities are somehow politically immune to the most fervent anti-migrant sentiments. On the contrary, as discussed in Sect. 6.6, European cities have long been incubators of the politics of racism and anti-immigration as well as testing grounds for restrictive government policies. The emergence of social media-savvy urban-based organizations such as PEGIDA in Germany, the racist pogroms that erupted in English cities in July 2024 and the anti-immigrant demonstrations outside asylum seeker hotels across the UK in August 2025, have made research on how the revival of nationalist and racist ideologies affects the relationship between migration and cities all the more urgent.

Meanwhile, urban social movements have increasingly placed migration at the heart of struggles for economic, racial, gender and environmental justice. The Black Lives Matter protests in 2020, for example, initiated a global public debate on how to address the memory of forced movement vis-à-vis public narratives of local history and identity. In Europe, this was epitomized by the toppling of the statue of seventeenth-century slave trader Edward Colston in Bristol and the subsequent petition to replace it with a monument to the 1963 Bristol Bus Boycott, when black

Caribbean activists successfully overturned the city's colour bar policy for public transport workers (Moody, 2021). Such events resonate with recent mobilizations around migration heritage (Sect. 5.4.2) but also reiterate the importance of urban public space as an arena for migration-related politics (Sect. 6.5).

Two major events in particular have raised important questions for research on the migration-city nexus. The Covid-19 pandemic severely disrupted mobility regimes and brutally exposed their uneven impact in terms of class, gender, race, citizenship and residency (Triandafyllidou, 2022). National lockdowns had the most vivid impact on cities. Restrictions on intra-urban mobility disproportionately affected low-income migrant households in small or overcrowded dwellings, resulting in prolonged periods of little or no income, increased school dropout rates and a heightened risk of infection (Manz & Plöger, 2024). The pandemic greatly reduced the circular migration of live-in care workers between workplaces in western European cities and home countries, revealing the fragility of transnational shift-work systems but also stranding workers in their clients' homes (Leiblfinger et al., 2020; Sect. 3.4). The pandemic also undermined dominant understandings of labour migration by elevating the status, if not the wages, of previously 'disposable' workers such as food delivery couriers into 'essential' workers, increasing their relative mobility but also their exposure to disease (Triandafyllidou, 2022: 83–84; Sect. 3.5). Popular geographical knowledge of urban diversity, as discussed in Sect. 4.1, was now used not to celebrate or denigrate multicultural urbanism, but to make false claims about the epicentres of Covid-19 outbreaks. Prato in Italy, for example, was publicly singled out in early 2020 as a likely hotspot due to its significant Chinese population, despite actually recording a low number of cases (Krause & Bressan, 2020). The pandemic thus exposed and exacerbated the differentiated mobilities within and between cities, the centrality of citizenship and residency in defining local belonging and access to services, and the underlying inequalities in work, health, education and housing. While a flurry of studies mapped the immediate impact of the virus in 2020, relatively little is known about the long-term consequences of Covid-19 for migrants across urban contexts (Fogelman & Cohen, 2024). Indeed, it often seems that the pandemic has become a passing reference (as is the case here!) or has simply disappeared from discussion.

Like the pandemic, the war in Ukraine has both interrupted and intensified pre-existing patterns of urban migration. In the first two months following the Russian invasion in February 2022, three million refugees left Ukraine for neighbouring countries (UNHCR, 2024). Cities in Central and Eastern Europe with limited experience of international migration, such as Warsaw, now found themselves at the forefront of managing hundreds of thousands of foreign newcomers (Sect. 6.3.3). At the same time, the war has highlighted the systemic racism embedded in European immigration policies, which now favoured Ukrainians over other nationals. For example, the Danish government waived its restrictions on the concentration of immigrants in social housing (see Sect. 4.3.4), allowing Ukrainian refugees to settle in accommodation that had been emptied of 'non-Westerners' (Boffey, 2022). The war has also tested Western liberal ideals of urban diversity. It is ironic that *Odessa*—a city founded on immigration (Sect. 2.1) and whose name has long stood as a metonym for internationalism—now appears in major Anglophone media outlets as *Odesa*, thus endorsing the Ukrainian nationalist name for the Black Sea port, albeit one under the intermittent onslaught of Russian missiles. It suggests that the violence of war not only displaces and exterminates particular groups in cities, but also unsettles our cosmopolitan imaginary. Beyond the conventional focus on refugee movements, war transforms the relationship between migration and cities in complex and unpredictable ways, both at the frontline and far beyond (Dines, 2025).

7.3 Some Pointers for Future Research

As the links between migration and cities continue to evolve, new questions will continue to emerge and gaps in existing knowledge will be identified. Important research will continue to be conducted on 'classic' topics such as labour markets, residential segregation and local policymaking, and new theoretical and methodological perspectives will be developed, perhaps inspired by work conducted in other fields and perhaps wrestling with some of the issues discussed in this book.

I conclude with some thoughts on areas for future research. Some have just been mentioned: the rise of the populist right, the long-term effects of the Covid-19 pandemic and the consequences of armed conflicts such as the war in Ukraine. In addition, each thematic chapter has identified promising new directions and questions that merit further investigation. The following brief list is not meant to be exhaustive, but rather offers a hint of the directions that work on the migration-city nexus can take.

In the field of history (Chapter 2), there is a growing body of exciting research that challenges schematic accounts of the periodization, types and origins of migration associated with particular cities. For example, by drawing on multiple sources, including oral histories, there is great scope to complicate knowledge of post-war labour migrations and thus dominant narratives about urbanization and migrant settlement, as Jozefien De Bock's (2018) work on Ghent demonstrates.

In terms of urban economies (Chapter 3), theoretical and empirical work on social reproduction and reproductive labour (Farris, 2020; Rigo, 2026), as well as critical debates on racial capitalism (Bhattacharyya, 2018; De Genova, 2023), have the potential to deeply transform our thinking on the relationship between migration and urban development, and can be productively brought into conversation with Glick Schiller and Çağlar's trailblazing research on urban rescaling.

On the question of migrant settlement processes (Chapter 4), there are numerous open questions that can be further explored through transnational comparative work and intersectional analysis, for example how urban newcomers are differently implicated in short-term rental markets as hosts, guests and service workers, and how this shapes their pathways of incorporation.

Cultural processes (Chapter 5) are increasingly recognized by migration scholars as an important focus in their own right (Martiniello, 2015). Heritage and public memory, in particular, represent compelling arenas for research, both as a means to engage with grassroots and institutional mobilizations around the meanings of urban diversity and the links between past and present migrations, and to address the contradictory relationship between heritage consumption and urban dynamics such as regeneration and gentrification (De Cesari & Dimova, 2019).

Finally, on the topic of urban governance, a rich and extensive body of policy analysis has been produced in recent years (Sect. 6.3.3), but we still know surprisingly little about the messy contexts in which policies are made and how they affect the everyday lives of migrants. Insights from the current literature would be greatly enhanced by 'behind-the-scenes' ethnographic research.

Other emerging issues, not covered in this book, are destined to become increasingly important in the near future. Two in particular are climate change and artificial intelligence. The main reason for omitting these two topics is that, although there is growing scholarly interest in their links to migration (Beduschi, 2021; Bettini et al., 2017), there is little evidence-based literature on the specific question of migration and European cities, while certain big-picture discussions remain somewhat speculative. Indeed, going forward, urban migration scholars will need to interrogate the usefulness of categories (see, for example, Bettini et al.'s (2017) critique of the terms 'resilience' and 'climate refugee') and the ways in which the impacts of both climate change and artificial intelligence are interwoven with existing social structures and inequalities (Bettini et al., 2021; Zajko, 2022).

Another absence in this book, acknowledged from the outset, is the multifarious urban world that exists outside Europe. There have been important moves over the last two decades to reformulate urban theory from global South perspectives (Sect. 1.3.2). An established literature has also considered the relationship between mobility and urbanization outside the West (Roy & Ong, 2011). As suggested in the book's introduction, recent debates on postcolonial urban theory and comparative urbanism (Robinson, 2006) can also be applied to address the internal differences and uneven knowledge production within Europe itself. Much work, however, remains to be done on developing comparative strategies that can meaningfully bring migration and urban processes across various Norths and Souths into conversation. As my own research has moved from Italy to consider cities in North Africa, I have found myself grappling with the connections and disconnections—urban informality, mass internal migration, colonialism and so on—that configure the migration-city nexus in a trans-Mediterranean space (Dines, 2022).

Ultimately, the underlying goal of research on migration and cities remains the same: to fathom how migration transforms cities and how the urban scale invites us to rethink migration. While new issues will inevitably arise, many of the fundamental questions explored in the various chapters will continue to demand our attention. Hopefully, this book has offered inspiration for moving forward.

References

Aalbers, M. (2018). Financial geography II: Financial geographies of housing and real estate. *Progress in Human Geography, 43*(2), 376–387.

Anderson, B. (2019). New directions in migration studies: towards methodological de-nationalism. *Comparative Migration Studies, 7*(36). https://doi.org/10.1186/s40878-019-0140-8

Arbaci, S. (2019). *Paradoxes of segregation: Housing systems, welfare regimes and ethnic residential change in southern European cities.* Wiley.

Beduschi, A. (2021). International migration management in the age of artificial intelligence. *Migration Studies, 9*(3), 576–596.

Bettini, G., Nash, S. L., & Gioli, G. (2017). One step forward, two steps back? The fading contours of (in)justice in competing discourses on climate migration. *The Geographical Journal, 183*(4), 348–358.

Bettini, G., Beuret, N., & Turhan, E. (2021). On the frontlines of fear: Migration and climate change in the local context of Sardinia, Italy. *ACME: An International Journal for Critical Geographies, 20*(3). https://doi.org/10.14288/acme.v20i3.1838

Bhattacharyya, G. (2018). *Rethinking racial capitalism: Questions of reproduction and survival.* Rowman & Littlefield.

Boffey, D. (2022, May 2). Denmark accused of racism after anti-ghetto law adapted for Ukrainians. *The Guardian.* https://www.theguardian.com/world/2022/may/02/denmark-accused-racism-anti-ghetto-law-ukraine-refugees

Crul, M. (2016). Super-diversity vs. assimilation: how complex diversity in majority-minority cities challenges the assumptions of assimilation. *Journal of Ethnic and Migration Studies, 42*(1), 54–68.

Dahinden, J. (2016). A plea for the "de-migranticization" of research on migration and integration. *Ethnic and Racial Studies, 39*(13), 2207–2225.

De Bock, J. (2018). *Parallel lives revisited: Mediterranean guest workers and their families at work in the neighbourhood, 1960–1980.* Berghahn.

De Cesari, C., & Dimova, R. (2019). Heritage, gentrification, participation: Remaking urban landscapes in the name of culture and historical preservation. *International Journal of Heritage Studies, 25*(9), 863–869.

De Genova, N. (2023). A racial theory of labour: Racial capitalism from colonial slavery to postcolonial migration. *Historical Materialism, 31*(3), 219–251.

Dines, N. (2016). From "southern" to "ordinary": Conceptualizing and contextualizing segregation in public space in Euro-Mediterranean cities. *Méditerranée, 127*, 101–108.

Dines, N., Montagna, N., & Vacchelli, E. (2018). Beyond crisis talk: Interrogating migration and crises in Europe. *Sociology, 52*(3), 439–447.

Dines, N. (2022). How far can we compare? Migration studies, comparative urbanism and the potential of a trans-Mediterranean perspective. *International Migration, 60*(1), 205–218.

Dines, N. (2025). War, cities and migration-driven diversity: Unravelling a complex relationship. In A. Devasundaram, Z. Holman, M. Mattos, & S. Zografos (Eds.), *Urban violence and marginalised communities: Multidisciplinary investigations.* UCL Press.

Farris, S. (2020). The business of care: Private placement agencies and female migrant workers in London. *Gender, Work & Organization, 27*(6), 1450–1467.

Fogelman, T., & Cohen, N. (2024). Pandemic urban citizenship: Cities, migrants, and Covid-19 crisis. Introduction to the special issue. *Cities, 158.* https://doi.org/10.1016/j.cities.2024.105676

Glick Schiller, N., & Çağlar, A. (2009). Towards a comparative theory of locality in migration studies: Migrant incorporation and city scale. *Journal of Ethnic and Migration Studies, 35*(2), 177–202.

Ha, N., & Picker, G. (Eds.). (2022). *European cities: Modernity, race and colonialism.* Manchester University Press.

Hall, S., Critcher, C., Jefferson, T., Clarke, J., & Roberts, B. (1978). *Policing the crisis: Mugging, the state, and law and order*. Macmillan.

Harris, R. (2024). A tale of four cities: Neighbourhood diversification and residential desegregation in and around England's 'no majority' cities. *The Geographical Journal, 190*(2). https://doi.org/10.1111/geoj.12561

Kaufmann, V. (2012). *Rethinking the city: Urban dynamics and motility*. Routledge.

Krause, E., & Bressan, M. (2020). Viral encounters: Xenophobia, solidarity, and place-based lessons from Chinese migrants in Italy. *Human Organization, 79*(4), 259–270.

Leiblfinger, M., Prieler, V., Schwiter, K., Steiner, J., Benazha, A., & Lutz, H. (2020, May 14). Impact of the COVID-19 pandemic on live-in care workers in Germany, Austria, and Switzerland. https://ltccovid.org/2020/05/14/impact-of-the-covid-19-pandemic-on-live-in-careworkers-in-germany-austria-and-switzerland/

Maloutas, T. (2012). Contextual diversity in gentrification research. *Critical Sociology, 38*(1), 33–48.

Maloutas, T., & Karadimitriou, N. (Eds.). (2022). *Vertical cities: Micro-segregation, social mix and urban housing markets*. Elgar.

Manz, M., & Plöger, J. (2024). Lockdown/locked out – impact of the pandemic on the unequal geography in German cities. *Cities, 158*. https://doi.org/10.1016/j.cities.2024.105171

Martiniello, M. (2015). Immigrants, ethnicized minorities and the arts: A relatively neglected research area. *Ethnic and Racial Studies, 38*(8), 1229–1235.

Moisio, S., & Rossi, U. (2024). *The urban field: Capital and governmentality in the age of techno-monopoly*. Columbia University Press.

Moody, J. (2021). Off the pedestal. The fall of Edward Colston. *Public History Review, 28*. https://doi.org/10.5130/phrj.v28i0.7776

Pisarevskaya, A., & Scholten, P. (2025). *Cities of migration. Understanding the diversity of urban diversities in Europe*. Springer.

Rigo, E. (2026). *Migration, asylum and exploitation: A feminist legal perspective*. Bristol University Press.

Robinson, J. (2006). *Ordinary cities: Between modernity and development*. Routledge.

Roitman, J. (2014). *Anti-crisis*. Duke University Press.

Roy, A., & Ong, A. (Eds.). (2011). *Worlding cities: Asian experiments and the art of being global*. Blackwell.

Srnicek, N. (2017). *Platform capitalism*. Polity Press.

Triandafyllidou, A. (2022). Migration and the pandemic emergency. In A. Triandafyllidou (Ed.), *Routledge handbook of immigration and refugee studies* (pp. 82–94). Routledge.

UNHCR. (2024, February 16). Ukraine refugee situation. https://data.unhcr.org/en/documents/details/106707

Wacquant, L. (2007). *Urban outcasts. A comparative sociology of advanced marginality*. Polity Press.

Walter, S. (2021). The backlash against globalization. *Annual Review of Political Science., 24*, 421–442.

Zajko, M. (2022). Artificial intelligence, algorithms, and social inequality: Sociological contributions to contemporary debates. *Sociological Compass, 16*(3). https://doi.org/10.1111/soc4.12962

MIX
Papier aus verantwortungsvollen Quellen
Paper from responsible sources
FSC® C105338

If you have any concerns about our products,
you can contact us on
ProductSafety@springernature.com

In case Publisher is established outside the EU,
the EU authorized representative is:
Springer Nature Customer Service Center GmbH
Europaplatz 3, 69115 Heidelberg, Germany

Printed by Libri Plureos GmbH
in Hamburg, Germany